Geology

2nd Edition

by Alecia M. Spooner

for dummies®

A Wiley Brand

Geology For Dummies®, 2nd Edition

Published by: **John Wiley & Sons, Inc.**, 111 River Street, Hoboken, NJ 07030-5774, www.wiley.com

Copyright © 2020 by John Wiley & Sons, Inc., Hoboken, New Jersey

Published simultaneously in Canada

For general information on our other products and services, please contact our Customer Care Department within the U.S. at 877-762-2974, outside the U.S. at 317-572-3993, or fax 317-572-4002. For technical support, please visit https://hub.wiley.com/community/support/dummies.

Wiley publishes in a variety of print and electronic formats and by print-on-demand. Some material included with standard print versions of this book may not be included in e-books or in print-on-demand. If this book refers to media such as a CD or DVD that is not included in the version you purchased, you may download this material at http://booksupport.wiley.com. For more information about Wiley products, visit www.wiley.com.

Library of Congress Control Number: 2020902701

ISBN 978-1-119-65287-8 (pbk); ISBN 978-1-119-65292-2 (ebk); ISBN 978-1-119-65291-5 (ebk)

Manufactured in the United States of America

SKY10030575_101521

Contents at a Glance

Table of Contents

Introduction

Geology is the study of the earth. By default this means that geology is a vast, complex, and intricate topic. But "vast, intricate, and complex" does not necessarily mean difficult. Many folks interested in geology just don't know where to start. Minerals? Rocks? Glaciers? Volcanoes? Fossils? Earthquakes? The sheer number of topics covered under the heading "geology" can be overwhelming.

Enter *Geology For Dummies!* The goal of this book is to break through the overwhelming array of geology information and provide a quick reference for key concepts in the study of the earth.

My hope is that you find this book both interesting and useful, whether you've purchased it to accompany a course you're taking in school or to help you find answers to questions you have about the planet you live on.

About This Book

In *Geology For Dummies,* you can start anywhere. This book is written as an introduction to the most common topics in geology. Follow your interest from one topic to the next, or start at the beginning and read the chapters in order. I wrote the book in a style that allows you to open to any page and learn something. But if you want to start at the beginning, you're introduced to the concepts in a logical and structured order that (I hope!) answers your questions almost as soon as you ask them.

Throughout the book you find cross-references to other chapters. I use them because it's impossible to explore one topic in geology without touching on many others. The multiple cross-references weave together the different parts of geologic study into a complex whole.

Wherever possible, I include illustrations to accompany my explanations. Geology is all around you, so while you are busy reading this book and examining the illustrations, I encourage you to also look around and find real-world examples of the processes and features I describe. To this end, I have also included a color photo section in the middle of the book featuring vivid images that help bring the subject matter to life.

Foolish Assumptions

As I was writing this book, I had to make a few assumptions about you, the reader. I assume that you live on Earth and are familiar with rocks, streams, and weather (rain, wind, and sun). I also assume that you are familiar with a very basic geography of the earth, including the continents, oceans, and major mountain ranges.

I do not assume that you have any scientific background in chemistry, which you may find useful if you want to dig deeper into the details of rock formation and transformation. Similarly, when I discuss evolution I do not assume that you have any background in biology or anatomy (and none is needed to understand the concepts I present). If the subject of evolution interests you, you may find that your questions lead you to pick up other reference books on that topic.

If you find that your interest in geology is further fueled by this book, I recommend that you purchase an earth science or geology dictionary. Geology is full of terms with precise and informative meanings. With this kind of dictionary on hand, you'll find you can easily interpret even the most befuddling geological explanations.

Icons Used in This Book

Throughout this book, I use icons to draw your attention to certain information:

TIP

The Tip icon indicates information that may be especially useful to you as you prepare for a geology exam or assignment or as you begin studying geology on your own.

WARNING

This icon, which appears only rarely in this book, points out situations that may be dangerous.

REMEMBER

Information highlighted with the Remember icon is foundational to understanding the concept being explained. Sometimes this icon indicates a definition or concise explanation. Other times it indicates information that will help you tie multiple concepts together.

TECHNICAL
STUFF

This icon indicates that the information goes a little beyond the surface into some technical details. These details are not necessary for your broad understanding of the topic or concepts, but you may find them interesting and informative.

Beyond the Book

In addition to what you're reading right now, this product also comes with a free access-anywhere Cheat Sheet that tells you about plate tectonics and the geologic timescale.. To get this Cheat Sheet, simply go to `www.dummies.com` and type **Geology For Dummies Cheat Sheet** in the Search box.

Where to Go from Here

You have most likely purchased this book with a question about geology already in mind. In that case, I encourage you to follow your interest. Use the table of contents or index to find where I answer your question, flip to that page, and get started!

If you don't have a particular question in mind, here are a few of my favorite topics that will get you started on your study of Earth:

>> **Chapter 8, "Adding Up the Evidence for Plate Tectonics":** In this chapter, I tell you the story of how an early geologist, Alfred Wegener, began to think about plate movements. He collected evidence to support his ideas, but it took many years before the idea of plate tectonics was accepted by the scientific community. This chapter is a great introduction to how science really happens, as well as an overview of the foundational theory of modern geology.

>> **Chapter 12, "Water: Above and Below Ground":** If you want to get started by reading about something you can relate to, start with flowing water. Streams and rivers are the most common geologic processes on Earth. Regardless of where you live, you have probably witnessed the action of flowing water moving sediment or rocks. This chapter provides details from how water picks up and carries particles, to how rivers carve canyons and caves. It also covers the topic of groundwater, which is where most of the water you drink comes from.

>> **Chapter 18, "Time before Time Began: The Precambrian":** Long ago in Earth's deep, dark, murky past lay the beginnings of life. This chapter describes the first few billion years of Earth's existence, from its formation from a gaseous cloud, up to and including the earliest evidence for life — in the form of trace fossils called *stromatolites*.

1

Studying the Earth

Discover you are already a scientist, asking questions and seeking answers every day!

Learn the history and development of geologic study.

Go on a guided tour of Earth's systems, from the atmosphere to the inner core and everything in between.

Chapter **1**

Rocks for Jocks (and Everybody Else)

eology and earth sciences seem to have a reputation for being easy sub-jects, or at least the least difficult of the science courses offered in high school and college. Perhaps that's because the items observed and studied in geology —rocks — can be held in your hand and seen without a microscope or telescope, and they can be found all around you, anywhere that you are.

However, exploring geology is not just for folks who want to avoid the heavy cal-culations of physics or the intense labs of chemistry. Geology is for everyone. Geology is the science of the planet you live on — the world you live in — and that is reason enough to want to know more about it. *Geology* is the study of the earth, what it's made of, and how it came to look the way it does. Studying geology means studying all the other sciences, at least a little bit. Aspects of chemistry, physics, and biology (just to name a few) are the foundation for understanding Earth's geologic system, both the processes and the results.

Finding Your Inner Scientist

You are already a scientist. Maybe you didn't realize this, but just by looking around and asking questions you behave just like a scientist. Sure, scientists call their approach of asking and answering questions the *scientific method*, but what you do every day is the very same thing, without the fancy name. In Chapter 2, I present the scientific method in detail. Here, I offer a quick overview of what it entails.

Making observations every day

Observations are simply information collected through your five senses. You could not move through the world without collecting information from your senses and making decisions based on that information.

Consider a simple example: Standing at a crosswalk, you look both ways to determine if a car is coming and if the approaching car is going slow enough for you to safely cross the street before it arrives. You have made an observation, collected information, and based a decision on that information — just like a scientist!

Jumping to conclusions

You constantly use your collected observations to draw conclusions about things. The more information you collect (the more observations you make), the more solid your conclusion will be. The same process occurs in scientific exploration. Scientists gather information through observations, develop an educated guess (called a *hypothesis*) about how something works, and then seek to test their educated guess through a series of experiments.

No scientist wants to jump to a false conclusion! Good science is based on many observations and is well-tested through repeated experiments. The most important scientific discoveries are usually based on the educated guesses, experiments, and continued questioning of a large number of scientists.

Focusing on Rock Formation and Transformation

As I explore in detail in Part 2 of this book, the foundation of geology is the examination and study of rocks. Rocks are, literally, the building blocks of the earth and its features (such as mountains, valleys, and volcanoes). The materials that make up rocks both inside and on the surface of the earth are constantly shifting from one form to another over long periods of time. This cycle and the processes of rock formation and change can be traced through observable characteristics of rocks found on Earth's surface today.

Understanding how rocks form

Characteristics of rocks such as shape, color, and location tell a story of how and where the rocks formed. A large part of geologic knowledge is built on understanding the processes and conditions of rock formation. For example, some rocks form under intense heat and pressure, deep within the earth. Other rocks form at the bottom of the ocean after years of compaction and cementation. The three basic rock types, which I discuss in detail in Chapter 7, are:

>> **Igneous:** Igneous rocks form as liquid rock material, called *magma* or *lava,* cools. Igneous rocks are most commonly associated with volcanoes.

>> **Sedimentary:** Most sedimentary rocks form by the cementation of sediment particles that have settled to the bottom of a body of water, such as an ocean or lake. (There are also some sedimentary rocks, which are not formed this way. I describe these in Chapter 7 as well.)

>> **Metamorphic:** Metamorphic rocks are the result of a sedimentary, igneous, or other metamorphic rock being squeezed under intense amounts of pressure or subjected to high amounts of heat (but not enough to melt it) that change its mineral composition.

Each rock exhibits characteristics that result from the specific process and environmental conditions (such as temperature, or water depth) of its formation. In this way, each rock provides clues to events that happened in Earth's past. Understanding the past helps us to understand the present and, perhaps, the future.

Tumbling through the rock cycle

The sequence of events that change a rock from one kind into another are organized into the rock cycle. It is a cycle because there is no real beginning or end. All the different types of rocks and the various earth processes that occur are included in the rock cycle. This cycle explains how materials are moved around and recycled into different forms on the earth's surface (and just below it). When you have a firm grasp on the rock cycle, you understand that every rock on Earth's surface is just in a different phase of transformation, and the same materials may one day be a very different rock!

Mapping Continental Movements

Most of the rock-forming processes of the rock cycle depend on forces of movement, heat, or burial. For example, building mountains requires force exerted in two directions, pushing rocks upward or folding them together. This type of movement is a result of continental plate movements. The idea that the surface of the earth is separated into different puzzle-like pieces that move around is a relatively new concept in earth sciences, called *plate tectonics theory* (the subject of Part 3).

Unifying geology with plate tectonics theory

For many decades, earth scientists studied different parts of the earth without knowing how all the features and processes they examined were tied together. The idea of plate movements came up early in the study of geology, but it took a while for all the persuasive evidence to be collected, as I describe in Chapter 8.

By the middle of the twentieth century, scientists had discovered the Mid-Atlantic Ridge and gathered information about the age of sea floor rocks across the ridge. With this evidence they proposed the theory of plate tectonics suggesting the earth's crust is broken into pieces, or plates. Where two plates touch and interact is called a *plate boundary*.

Exactly how the earth's crustal plates interact is determined by the type of motion and type of crustal material. These interactions are described as plate boundary types and include:

>> **Convergent boundaries:** At convergent boundaries, two crustal plates are moving toward one another and come together. Depending on the density of the crustal plates, this collision builds mountains, or causes plate *subduction* (meaning one plate goes beneath another), producing volcanoes.

>> **Divergent boundaries:** At divergent plate boundaries, two crustal plates are separating or moving apart from one another. These boundaries are most commonly observed along the sea floor, where the upwelling of magma along the boundary creates a mid-ocean ridge, but they may also occur on continents, such as in the African rift valley.

>> **Transform boundaries:** At transform boundaries, the two plates are neither colliding nor separating; they are simply sliding alongside one another.

In Chapter 9, I provide the details on the different characteristics of continental plates and how they interact as they move around Earth's surface, including the particular geologic features associated with each plate boundary type.

Debating a mechanism for plate movements

While the unifying theory of plate tectonics has been well-accepted by the scientific community, geologists have yet to agree on what, exactly, drives the movement of continental plates.

Three dominant forces are thought to work together to drive plate tectonic motion:

>> **Mantle convection:** The *convection* of the mantle — the movement of heated rock materials beneath Earth's crust is thought to be the dominant driver of plate motion. Mantle rock moves outward towards the crust when it is heated, and then cools and sinks back towards the core (sort of like the wax in a lava lamp). As it moves, the crustal plates resting on the outer mantle are carried along.

>> **Ridge-push:** The ridge-push force is a result of new crustal rock forming at a mid-ocean ridge. The addition of new crust at the plate edge will push the plate away from the ridge and towards the plate boundary along its outer or opposite edge.

>> **Slab-pull:** As the ridge is pushing the plate away, the outer edge of such a plate will be sinking into the mantle, and as this slab sinks, it pulls the plate along behind it — creating the slab-pull force.

Mantle convection, ridge-push, and slab-pull forces work together to drive plate tectonic motion. In Chapter 10, you will find more details on how these three forces are constantly reshaping the surface of the earth.

Moving Rocks around on Earth's Surface

On a smaller than global scale, rocks are constantly being moved around on Earth's surface. Surface processes in geology include changes due to gravity, water, ice, wind, and waves. These forces sculpt Earth's surface, creating landforms and landscapes in ways that are much easier to observe than the more expansive processes of rock formation and tectonic movement. Surface processes are also the geologic processes humans are more likely to encounter in their daily lives.

>> **Gravity:** Living on Earth you may take gravity for granted, but it is a powerful force for moving rocks and sediment. Landslides, for example, result when gravity wins over friction and pulls materials downward. The result of gravity's pull is *mass wasting,* which I explain in Chapter 11.

>> **Water:** The most common surface processes include the movement of rocks and sediment by flowing water in river and stream channels. The water makes its way across Earth's surface, removing and depositing sediment, reshaping the landscape as it does. The different ways flowing water shapes the land are described in Chapter 12.

>> **Ice:** Similar to flowing water but much more powerful, ice moves rocks and can shape the landscape of an entire continent through glacier erosion and deposition. The slow-flowing movement of ice and its effect on the landscape are described in Chapter 13.

>> **Wind:** The force of wind is most common in dry regions, and you are probably familiar with the landforms it creates, called *dunes.* You may not realize that the speed and direction of wind create many different types of dunes, which I describe in Chapter 14.

>> **Waves:** Along the coast, water in the form of waves is responsible for shaping shorelines and creating (or destroying) beaches. In Chapter 15, I describe in detail the various coastal landforms created as waves remove or leave behind sediments.

Interpreting a Long History of Life on Earth

One of the advantages of studying geology is being able to learn what mysteries of the past are hidden in the rocks. Sedimentary rocks, formed layer by layer over long periods of time, tell the story of Earth's living history: changing climates and environments, as well as the evolution of life from single cells to modern complexity.

Using relative versus absolute dating

Scientists use two approaches to determine the age of rocks and rock layers: relative dating and absolute dating.

Relative dating provides ages of rock layers in relation to one another — for example, stating that one layer is older or younger than another is. The study of rock layers, or *strata*, is called *stratigraphy*. In methods of relative dating, geologists apply *principles of stratigraphy* such as these:

» Rock layers below are generally older than rock layers above.

» All sedimentary rock layers are originally formed in a horizontal position.

» When a different rock is cutting through layers of rock, the cross-cutting rock is younger than the layers it cuts through.

These principles and a few others that I describe in Chapter 16 guide geologists called *stratigraphers* in interpreting the order of rock layers so that they can form a relative order of events in Earth's history.

However, sometimes simply knowing that something is older than — or younger than — something else is not enough to answer the question being asked. *Absolute dating* methods use radioactive atoms called *isotopes* to determine the age in numerical years of some rocks and rock layers. Absolute dating methods may determine, for example, that certain rocks are 2.6 million years old. These methods are based on the knowledge, learned from laboratory experiments, that some atoms transform into different atoms at a set rate over time. By measuring these rates of change in a lab, scientists can then measure the amount of the different atoms in a rock and provide a fairly accurate age for its formation.

If the process of obtaining absolute dates from isotopes seems very complex, don't worry: In Chapter 16, I explain in much more detail how absolute dates are calculated and how they are combined with relative dates to construct the *geologic timescale:* a sequence of Earth's geological history separated into different spans of time (such as periods, epochs, and eons).

Witnessing evolution in the fossil record

The most fascinating story told in the rock layers is the story of Earth's evolution. To *evolve* simply means to change over time. And indeed, the earth has evolved in the 4.5 billion years since it formed.

Both the earth itself and the organisms that live on Earth have changed through time. In Chapter 17, I briefly explain the biological understanding of evolution. Much of modern understanding about how species have changed through time is built on evidence from *fossilized* or preserved life forms in the rock layers.

Fossilization occurs through different geologic and chemical processes, but all fossils can be described as one of two forms:

>> **Body fossils:** Remains of an organism itself, or an imprint, cast, or impression of the organism's body.

>> **Trace fossils:** Remains of an organism's activity, such as movement (a footprint) or lifestyle (a burrow) but without any indication of the organism's actual body.

Earth did not always support life. In Chapter 18, I describe the very early Earth as a lifeless, hot, atmosphere-free planet in the early years of the solar system's formation. It took billions of years before simple, single-celled organisms appeared, and their origins are still a scientific mystery.

Simple, single-celled life ruled Earth for many millions of years before more complex organisms evolved. Even then, millions of years passed with soft-bodied life forms that are difficult to find in the fossil record. It wasn't until 520 million years ago that the *Cambrian explosion* occurred. Chapter 19 describes this sudden appearance of shell-building, complex life as well as the millions of years that followed when life was lived almost entirely in the oceans until amphibians emerged on the land.

Chapter 20 delves into the Age of Reptiles, when dinosaurs ruled the earth and reptiles filled the skies and seas. During this period, all the earth's continents were connected as Pangaea, Earth's most recent supercontinent. But before the Age of Reptiles ended, Pangaea broke apart into the separate continents you recognize today. Evidence for Pangaea is still visible in the coastal outlines of South America and Africa — indicating where they used to be attached as part of the supercontinent.

In relatively recent time, geologically speaking, mammals took over from reptiles to rule the earth. The Cenozoic era (beginning 65.5 million years ago), which we are still experiencing, is the most recent and therefore most detailed portion of Earth's history that can be studied in the geologic record (the rocks). Many of the most dramatic geologic features of the modern Earth, such as the Grand Canyon and the Himalayan Mountains, were formed in this most recent era. In Chapter 21, I describe the evolution of mammal species (including humans) and the geologic changes that occurred to bring us to today.

At various times in the history of Earth, many different species have disappeared in what scientists call *mass extinction events*. In Chapter 22, I describe the five most dramatic extinction events in Earth's history. I also explain a few of the common hypotheses for mass extinctions, including climate change and asteroid impacts. Finally, I explain how the earth may be experiencing a modern-day mass extinction due to human activity.

IN THIS CHAPTER

» **Finding your inner scientist**

» **Applying the scientific method**

» **Distinguishing scientific laws from scientific theories**

» **Understanding the language of geology**

Chapter **2**

Observing Earth through a Scientific Lens

G eology is one of many sciences that study the natural world. Before moving on to the details of geologic science, I want to spend a little time sorting out what exactly science is and does. In this chapter, I describe the elements of science and the scientific method, and I explain how you do science every day perhaps without even realizing it!

Realizing That Science Is Not Just for Scientists

Science is not a secret society for people who like to wear lab coats and spend hours looking into microscopes. Science is simply the asking and answering of questions. Any time you make a decision by considering what you know, collecting new information, forming an educated guess, and figuring out whether your guess is right, you participate in acts of science.

Take a very simple example: choosing a shampoo. You've probably tried different types or brands of shampoo, observed how each one leaves your hair looking and feeling, and then decided which shampoo you wanted to purchase the next time. This process of observation, testing, and decision-making is all part of the scientific approach to problem-solving. You follow this process every day in multiple situations as you make decisions about what to buy, what route to take in your car, what to eat for dinner, and so on.

REMEMBER

Don't underestimate the role of science in your daily life. Every interaction you participate in — with the physical world and with other people — is governed by the natural laws discovered and described by scientists in multiple fields of specialization. New products and technologies are the result of ongoing answer-seeking in the sciences. And explanations of how human beings effect and are effected by the natural world are constantly being updated by new scientific discoveries. Keep reading to find out how science is done using a step-by-step approach called the *scientific method.*

Using a Methodical Approach: The Scientific Method

Scientists seek to answer questions using a sequence of steps commonly called the *scientific method.* The scientific method is simply a procedure for organizing observations, making educated guesses, and collecting new information. The scientific method can be summarized as the following steps:

1. **Ask a question.** Scientists begin by asking, "Why does that happen?" or "How does that work?" Any question can be the start of your scientific journey. For example, "Why are my socks, which used to be white, now colored pink?"

2. **Form a hypothesis that answers your question.** A *hypothesis* is a proposed answer to your question: an educated guess based on what you already know. In science, a hypothesis must be testable, meaning that you (or someone else) must be able to determine if the hypothesis is true or false through an experiment. For example, "I think my socks turned pink because I washed them with pink laundry soap."

3. **State a prediction based on your hypothesis that can be tested.** Using the proposed explanation in your hypothesis, form a prediction that you can test. For example, "I predict that if I wash a white T-shirt with pink laundry soap, it will turn from white to pink."

4. **Design an experiment to test your prediction.** A good experiment is designed to best answer your question (see the upcoming "Testing your hypothesis: Experiments" section) by controlling as many factors as possible.

For example, to test the above prediction, I will wash one white T-shirt with white laundry soap and one white T-shirt with pink laundry soap. I will wash them in the same washing machine with the same type of water so that everything (except the soap) is the same.

5. **Perform the experiment.** Time to do the laundry! If my prediction is correct, the shirt washed with pink soap will turn pink.

6. **Observe the outcome.** Both white T-shirts are still white after being washed with the different types of soap.

7. **Interpret and draw conclusions from the outcome of the experiment.** Scientists may run a single experiment multiple times in order to get as much information as possible and make sure that they haven't made any mistakes that could affect the outcome. After they have all this new information, they draw a conclusion. For example, in my experiment, it appears that the color of the laundry soap is not what turns white T-shirts pink in the washing machine. At this point I can propose a new hypothesis about why my T-shirts have turned pink, and I can conduct a new experiment.

REMEMBER

8. **Share the findings with other scientists.** This is possibly the most important step in the process of science. Sharing your results with other scientists provides you with new insights to your questions and conclusions. In my example, I did not confirm my hypothesis. Quite the opposite: I confirmed that the color of the laundry soap is *not* responsible for changing the color of my T-shirts. This is still very important information for the community of scientists trying to determine what, exactly, turns white T-shirts pink in the washing machine. Knowing my results will lead other scientists to develop and test new hypotheses and predictions.

Next, I describe in more detail each step of the scientific method approach to answering questions.

Sensing something new

The first step in the scientific method is simply to use your senses. What do you see, feel, taste, smell, or hear? Each of your senses helps you collect information or *observations* of the world around you. Scientific observations are information collected about the physical world without manipulating it. (Manipulations come later, with experiments; keep reading for the details!)

After you have collected multiple observations, you may find that there is a pattern — each dog you pet feels soft — or you may find that some observations are different from the others — most of the dogs have brown fur, but some have white fur with black spots. By summarizing your observations in this way, you prepare to take the next step in the scientific method, developing a hypothesis.

I have a hypothesis!

After you have summarized your observations, it's time to propose an educated guess about the processes behind the patterns you observe. That educated guess is your *hypothesis*. In everyday speech people often say, "I have a theory" when they really mean "I have a hypothesis." (I'll get to theories in a few pages.)

REMEMBER

A hypothesis is an inference about the patterns you have observed, based on your observations and any previous knowledge you have about the topic. It's possible to have many different hypotheses about the observed patterns. How do you know which one is correct? You test it with an experiment, which I describe next.

Testing your hypothesis: Experiments

Now the real fun begins: experimenting to determine if one of your hypotheses is correct. Scientists use their hypotheses to develop predictions that can be tested. Based on the observations about the color of dog fur, a prediction could be this: "I predict that all dogs have either brown fur or white fur with black spots." The prediction is a restatement of my hypothesis, based on my observations.

To determine if my prediction is correct I need to collect more information. I will make new observations, but this time I will manipulate the situation and observe the outcome. In other words, this time my observations will be the result of an *experiment.*

REMEMBER

In science, the *experimental design*, or the way you go about collecting the new information, is very important. An experimental design describes the parameters of your experiment: how many samples you will take (how many observations you will make) and how you will choose those samples. These decisions are partly determined by the question you are asking and partly determined by the nature of the observations you are collecting.

In most cases it's impossible to observe every single instance of the physical world that you are exploring. Therefore, you must take a sample that can represent the rest. For example, I can't look at every dog in the world to see what color their fur is, so I may decide that looking at 100 dogs will provide me with enough observations to determine if my prediction is correct. Those 100 are a sample of the worldwide population of dogs. If I choose those 100 dogs wisely, they may be a very accurate representation of the worldwide dog population. The best sample size is different for each experiment; it all depends on the question being asked.

In earth science, experiments are often *natural* or *observational.* This means that scientists go out into the field and observe events that have already happened, such as the formation of rocks, rock layers, or features of the landscape. Scientists make these observations without changing any aspect of the event or its result.

Geologists also use another kind of experiment called a *manipulative* experiment. A manipulative experiment is done in a laboratory, where the scientist can manipulate or change certain factors in order to test which factors are most important in creating the observed outcome. In this case, multiple experiments can be done, each one testing the importance of a different factor (or variable), with the goal of zeroing in on the one (or ones) that explain the observed outcome.

Most importantly, a scientific experiment, whether it is a natural or manipulated experiment, must be repeatable. This means that the scientists must clearly describe the steps they have taken so that another scientist can repeat the same experiment and see if she too, gets the same result.

Crunching the numbers

After running experiments and making observations, a scientist is left with a large collection of information, or *data,* to use to draw a conclusion. Trying to find patterns in page after page of descriptive observations or lists of numbers is almost impossible. To find patterns in the data, a scientist uses statistics.

Statistics are a mathematical tool for describing and comparing information (observations) *quantitatively,* which simply means using numbers. By using numbers to describe the data, such as the number of times a certain characteristic is observed in different rock samples, scientists can organize and compare the patterns in the data using simple arithmetic.

Some people find statistics intimidating because they seem like complicated mathematical formulas. But really, statistical methods are simple mathematics combined in a step-by-step sequence to uncover patterns in the data. Some statistics determine if two sets of data have overall similarities or differences. Others determine which variables are most important in creating the observed outcomes.

Another reason scientists organize and describe their data quantitatively is so that they can display it using graphs. Many different types of graphs are used, and a scientist must determine which type of graph best displays the data in an understandable way. The most suitable graph depends on what type of data is being displayed. Figure 2-1 illustrates a few common graph types used in earth science:

>> **Pie graph:** This type of graph is best used for illustrating different pieces of a whole. The total of a pie chart must always add up to 100 percent.

>> **Bar graph:** Also called *histograms,* bar charts are used to display information that can be sorted into different categories.

>> **Scatterplot:** Scatterplot graphs illustrate how two types of data are related. Sometimes a scientist will use a scatterplot to look for patterns of relationship between the data types — by finding clusters of data points.

>> **Line graph:** This type of graph is most commonly used to plot changes in a type of data over time, distance, or other variable.

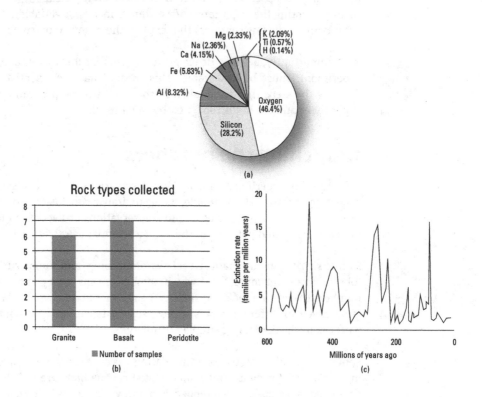

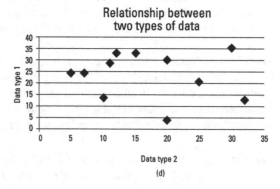

FIGURE 2-1:
a) Pie graph;
b) Bar graph;
c) Line graph;
d) Scatterplot.

Interpreting results

After data has been described, compared, and graphed, the next step is interpreting the data to draw new conclusions and perhaps propose a new hypothesis for further testing. Often scientists will find that the patterns in their data bring up new questions for exploration.

REMEMBER

If an experiment is designed well, the outcome (and collected data) should clearly prove or disprove the initial hypothesis. It is much easier and more common for a scientist to prove a hypothesis wrong than to prove it right. Finding that the hypothesis is incorrect helps rule out wrong ideas and is a very important step toward eventually finding an answer to larger questions that are being asked — and toward determining which hypothesis to test next.

The challenge at this stage is applying previous knowledge (perhaps from previous experiments) to understand what the patterns in the data — or the relationships between variables — mean. Rather than finding answers to all the questions, scientists often find themselves asking new questions and circling back to the hypothesis stage, preparing to test another hypothesis.

Sharing the findings

When a scientist has completed experiments, analyzed data, and interpreted the results, he must share his findings and ideas with other scientists. Commonly this step is done through scientific journals that are *peer-reviewed*, meaning that other qualified and respected scientists have examined the experimental design and procedure, perhaps tested it themselves, and determined that the results and interpretation are reasonable.

The peer-review step is very important. The process of having other knowledgeable scientists — other specialists in a particular topic — double-check the work helps find any errors. Errors may lead to false results or incorrect interpretations. Having more than one eye look for errors reduces the potential for moving forward on such false assumptions.

Building New Knowledge: A Scientific Theory

The goal of scientific study is to better understand the world. Step by step, information is collected until a broader or deeper understanding is gained. Eventually, this understanding may be expressed as a *scientific theory*. As scientists create and share theories, they expand what we know about the world around us.

It's never "just a theory"

Most people use the word *theory* to refer to an educated guess — a hypothesis. But scientifically speaking, a theory explains how some complex process works in the natural world. For example, the theory of plate tectonics that I cover in detail in Chapter 10 explains how crustal plates on the earth move around, forming mountains and volcanoes and causing earthquakes. The theory explains how all those geologic processes and resulting features are related to one another through the movement of crustal plates.

REMEMBER

A theory does not, however, explain *why* something occurs. The theory of plate tectonics does not answer the question of why the surface of the earth is broken into plates that move around. It only describes *how* those plates move around and interact with one another to result in the features we observe.

When a scientist describes something as a theory, she has come to the end of a long series of experiments and hypothesis testing. She is able to explain something so well, to provide evidence for that explanation (and to have that something accepted by other scientists as true) that it can be called a theory.

REMEMBER

In other words, a theory is a hypothesis that has been thoroughly tested through multiple experiments and is accepted as true by the scientific community. But the work doesn't end there! Scientists will continue to test hypotheses about the details within a theory, filling in gaps in understanding and looking out for incorrect assumptions that can be corrected to strengthen the theory.

Scientific theory versus scientific law

Scientific theories are not waiting to blossom someday into scientific laws. Laws and theories in science are two very different things.

>> **A scientific law describes an observed action that, when repeated many times, is always the same.** For example the *law of gravity* states that two objects will move toward one another. This movement is observed every time you drop something. The object you drop is attracted to the earth. The law of gravity simply describes this action, which is demonstrated to be the same in every test.

>> **A scientific theory explains how a set of observations are related.** For example, the *theory of gravity* seeks to explain how the relationship of two objects (their relative size, weight, and distance from each other) results in the observed interaction described by the law of gravity.

Both a scientific law and a scientific theory could be accurately described as "fact." Both are developed out of hypotheses that have been tested and proved true. A well-tested and generally accepted theory is considered true even though it may still be tested by the proposal of new hypotheses and experimentation. In some cases, part of a theory may be shown to be untrue, in which case the theory will be adjusted to accommodate this new truth without the entire theory being called into question.

The road to paradigms

A really thorough, well-tested, and widely accepted theory may become the current scientific paradigm. *Scientific paradigms* are patterns that serve as models for further research. Right now, plate tectonics theory is the paradigm within which all new geologic research takes place. The explanation provided by plate tectonics theory is accepted as proven true, and most researchers seek to answer questions that refine their understanding of this process rather than seeking to disprove the theory as a whole.

Paradigms, like theories, may change with new information; the change is called a *paradigm shift*. A paradigm shift brings a new perspective — a whole new way of looking at things. For example, the acceptance of the ancient age of the earth was a paradigm shift for early geologists. These scientists had struggled to explain how geologic features were created in the short span of a few thousand years (previously accepted as the age of the earth). The new paradigm of Earth being billions of years old provided a framework within which geologic processes had plenty of time to occur, creating the features they observed. (See Chapter 3 for more discussion about this particular paradigm shift.)

Speaking in Tongues: Why Geologists Seem to Speak a Separate Language

As with many sciences, when you begin to study geology you may find yourself overwhelmed by all the new words that you have to learn. Indeed, geologists have their own language for describing rocks, earth processes, and geologic features. But after you get the hang of what all the different words indicate, reading about geology is much less intimidating.

Lamination vs. foliation: Similar outcomes from different processes

Geologists describe characteristics of rocks with the intention of understanding the processes that formed those characteristics. For this reason, an observed characteristic, such as "layers," will have different terms indicating what kind of process resulted in those layers. Here's an example of what I mean:

>> A rock with layers may be described as *laminated*. Laminations are thin layers formed by the accumulation of tiny particles that settle through standing water (such as at the bottom of a lake or pond). This layered (laminated) rock is a sedimentary rock.

>> A rock with layers may also be described as *foliated*. Foliations are thin layers or sheets of minerals that are created by intense amounts of pressure and heat deep within the earth's crust. This layered (foliated) rock is a metamorphic rock.

REMEMBER

The layered characteristic of these rocks may seem similar at first glance but is actually the result of very different processes that occur under very different conditions on the earth. Closer inspection of the rocks (perhaps with a microscope) will reveal that the layers made of particles are different in appearance than the layers made of mineral sheets.

Gabbro vs. basalt: Different outcomes from similar processes

Another defining characteristic of rocks is their *composition*, or what minerals they are made of. However, rocks with the same mineral composition may have different names. Why? Geologists want to categorize rocks according to both composition *and* formation process. An example is the distinction between the rocks called *gabbro* and *basalt*.

Both gabbro and basalt are dark-colored rocks with the same mineral composition. They both are formed by the cooling of liquid rock (magma or lava) into a solid. Gabbro is formed when the liquid rock cools underground, slowly, over a long period of time. Basalt is formed when liquid rock cools very quickly, at or near the surface of the earth where it is exposed to air or water.

The different formation processes create observable differences in the rocks —
gabbro has large, visible mineral crystals, while basalt does not — but the
composition of the rocks is still the same. By giving them different names,
geologists can categorize a rock with a term that includes information about both
composition and formation characteristics.

TIP

When studying geology it is very helpful to have a dictionary of geologic terms or
dictionary of earth science handy to help tackle the immense amount of new
vocabulary that you will encounter!

IN THIS CHAPTER

» Using catastrophe to explain geologic phenomena

» Proposing origins for Earth's rocks

» Arriving at modern ideas about the earth

» Using today's processes to understand the past

» Unifying theories with plate tectonics

» Continuing to ask questions

Chapter **3**

From Here to Eternity: The Past, Present, and Future of Geologic Thought

For many sciences, the foundations of modern thought were laid during Europe's Scientific Revolution of the sixteenth and seventeenth centuries. During this time great thinkers began to redefine how they examined and understood the world around them. Although important advances were made in astronomy, mathematics, anatomy, and other sciences during this time, advancement of geologic science was constrained by a widely held belief that the Bible described an accurate age of the earth at only a few thousand years.

As a result, the road to modern geologic theories did not begin to be paved until later in the eighteenth and nineteenth centuries. In fact, significant insights into the earth's systems are still occurring today. In this chapter, I describe the important theories presented along the way to our current understanding of the earth and its systems. I also describe the sequence of important hypotheses that led to a unifying theory of the earth (called *plate tectonics* theory), as well as some of the exciting areas of current geologic research.

Catastrophe Strikes Again and Again

When early geologists looked at the mountains, valleys, and seas around them, they realized that something dramatic must have occurred to create what they saw. Because people believed that the earth was only a few thousand years old, the only way to explain what they saw was by assuming the occurrence of occasional catastrophic events, such as massive floods, volcanic eruptions, and earthquakes.

REMEMBER

This early belief that earth's features were created by a series of catastrophic events is called *catastrophism*.

Geologic explanations involving dramatic, worldwide, catastrophic events were in sync with stories from the Bible, such as the great flood. In this way, catastrophism reconciled strong biblical beliefs with explanations of geologic processes that scientists now know occur over many hundreds of thousands (or even millions) of years.

Early Thoughts on the Origin of Rocks

While catastrophism attempted to explain the creation of earth's features, questions about the origin of earth's rocks remained. Where did the rocks on the crust of the earth come from before they were subjected to the catastrophes that shaped and shifted them?

REMEMBER

Two theories dominated early thoughts on the origin of rocks: Neptunism and Plutonism.

>> **Neptunists proposed oceanic origins.** The Neptunist theory of rock origins proposed that all the rocks on Earth were created from sea water, having crystallized from the earth's first oceans. (The theory is named after the Roman god of the sea, Neptune.)

>> **Plutonists proposed volcanic origins.** The Plutonists believed that all Earth's rocks originated from volcanoes and were then changed by pressure and heat into other rocks. (The theory is named after the Roman god of the underworld, Pluto.)

TIP

While neither of these theories accurately explains how all rocks are formed, each contains partial truths. Some rocks do precipitate from ocean water, some rocks do form from volcanoes, and many rocks are changed into other rocks through heat and pressure, as you discover in Chapter 7.

Developing Modern Geologic Understanding

In this section, I introduce three of the most prominent geologists of the seventeenth and eighteenth centuries and describe the theories they proposed: theories that stand up to current scientific scrutiny and still form the basis for modern geologic understanding.

Reading the rock layers: Steno's stratigraphy

In the mid-seventeenth century, Nicholas Steno, a Danish physician, made great contributions to geology and especially *paleontology:* the study of fossil life. When Steno began his observations, only a few other scientists had proposed, tested, and attempted to prove that fossils found in rocks were the remains of once-living organisms. Steno advanced these ideas through observations and the study of rocks. His work led him to other questions, such as how could any solid object (a rock, mineral, or fossil) become trapped within another solid object, such as a rock?

REMEMBER

Steno is considered the father of modern *stratigraphy,* which is the study of layers of rock. He described four principles of stratigraphy that still hold true today:

>> **Principle of superposition:** States that in an uninterrupted sequence of *sedimentary* rocks (those composed of pieces of other rocks; see Chapter 7), the rock layers below are older than the rock layers above (as long as they have not been *deformed,* which I describe in Chapter 9). In Figure 3-1, the principle of superposition indicates that layer A is older than layer B, C, or E.

>> **Principle of original horizontality:** States that sediments forming sedimentary rocks are usually laid down in a horizontal position (due to gravity). Therefore, rock layers that appear vertical have been moved from their original, horizontal position by some natural force (such as an earthquake).

>> **Principle of lateral continuity:** States that when sediments are laid down, creating sedimentary rocks, they spread out until they reach some other object that confines them. This principle is illustrated when you fill your bathtub with water. The water spreads to fill all the space, confined only by the edges of the tub. Pour that same amount of water on the bathroom floor, and it spreads out until it hits the bathroom walls. Sedimentary rocks, like water, continue laterally until they are stopped by some other object.

>> **Principle of cross-cutting relationships:** States that where one type of rock cuts across or through another type of rock, the rock being cut is older and the rock cutting through is younger. After all, a rock must already exist in order to be cut through by another rock. This principle is illustrated in Figure 3-1 where rock unit D is younger than the rocks A, B, and C that it cuts through.

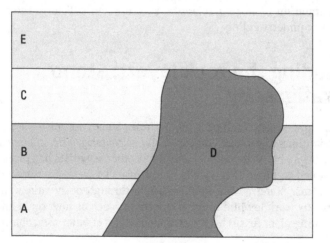

FIGURE 3-1: In this sketch of rock layers, the oldest is A, and the youngest is E.

These things take time! Hutton's hypothesis

While Plutonists and Neptunists were arguing about the origin of Earth's rocks, Scottish physician James Hutton was observing the rocks in his native Scotland and thinking about the different rock types and rock layers. He proposed theories about their relationship to one another and how current rock formations came to be.

FINDING SHARKS' TEETH ON MOUNTAINTOPS

Nicholas Steno scientifically examined the head of a shark and reported that the teeth in the shark's head were very similar to stones found in the mountains. These stones were called *glossopetrae,* or "tongue stones," and at the time were believed to result from lightning strikes. Indeed, what Steno had discovered was that sharks' teeth were actually buried in the rocks on the mountaintop!

REMEMBER

Through his observations of sites such as Siccar Point — a rocky area along the east coast of Scotland — Hutton began describing geologic processes that required long periods of time to create the rock formations visible today. He was the first geologist to propose the idea of geologic time, also called *deep time,* which extended the age of the earth much farther into the past than had previously been accepted.

According to Hutton, with a long enough period of time, even the small, commonplace processes that shape the earth's surface today could result in the dramatic formations previously assumed to be the results of catastrophe.

What has been will be: Lyell's principles

Following Hutton's work, Charles Lyell, a Scottish professor of geology in the early nineteenth century, published a book called *Principles of Geology.* In this book Lyell outlined and expanded on Hutton's ideas about deep time, geologic processes, and the formation of rock features on Earth's surface.

In publishing his book, Lyell spread Hutton's ideas and popularized them. The concept that "the present is the key to the past" was groundbreaking at the time and inspired scientific thought in fields outside of geology, such as Darwin's ideas about evolution.

REMEMBER

The basic principle that Hutton proposed, called *uniformitarianism,* is still the foundation of geologic science. Simply put, it states that past geologic phenomena can be explained by drawing on observable processes occurring today.

Uniformi-what? Understanding the Earth through Uniformitarianism

The idea that geologic processes we observe today have always been occurring and can be used to explain the features of the earth has stood the test of time. In fact, now more than ever, geologists recognize that the physical, chemical, and biological processes that occur today must have occurred in the past as well. Even a feature as spectacular as the Grand Canyon is created by the same simple process of erosion by water (see Chapter 12) that creates creeks and gullies in your backyard.

However, when Hutton and Lyell proposed the concept of uniformitarianism, they assumed that the rate and intensity of past processes were the same as those observed today. The current understanding of uniformitarianism in geology no longer makes this assumption. Modern uniformitarianism differs from the original idea in two very important ways:

» **Rates and intensity of processes may vary:** While the processes scientists observe today occurred in the past, they may have occurred more quickly or more intensely than they do now. For example, massive layers of volcanic rocks across Siberia (called the *Siberian Traps*) suggest a period of very intense lava outpourings, unlike anything humans have ever observed.

» **Catastrophes do play a role:** When uniformitarianism was first proposed, it ran counter to the ideas of catastrophism. But modern geologists recognize that occasional catastrophic events (such as volcanic eruptions and tsunamis) do play an important role in shaping the earth's surface.

Pulling It All Together: The Theory of Plate Tectonics

During World War I, a German scientist named Alfred Wegener suggested that the continents had once been connected and had drifted apart. His ideas about *continental drift* — the movement of the continents — were based on fossil, rock, and stratigraphic evidence (which I discuss in detail in Chapter 8). However, he hadn't worked out all the details – such as what force, or *mechanism*, propelled the continents. At the time, scientific understanding of the earth's crust as a continuous, solid, rigid layer did not allow for moving continents, and without a clear explanation for how they moved, Wegener's hypothesis was strongly rejected by other geologists.

A dramatic breakthrough occurred in the decades after World War I when the use of submarines in warfare led to mapping of the seafloor with sonar. The 1960s in particular was a time of new discovery and understanding. Geologist Marie Tharp discovered a long, rocky ridge, a *rift*, in the middle of the Atlantic Ocean while drawing an ocean floor map from sonar data. (Her map, created with Bruce Heezen, is still the standard map of the ocean floor used by all scientists.)

The idea of *seafloor spreading* — the moving apart of oceanic crust along ridges on the ocean floor — was being explored by scientists, led by the ideas of Harry Hess. But there were still many who "knew" the earth's crust and the mantle below were solid rock and could not be moving. One skeptic, Canadian geologist J. Tuzo Wilson, eventually published his ideas about plates moving across hotspots (see Chapter 10 for details) and his ideas provided a foundation on which the unifying theory of geology was built. The *theory of plate tectonics* combines ideas about plate movement with evidence for seafloor spreading, as well as incorporating explanations for volcanoes, earthquakes, and other geologic features and phenomena. Because this theory is so crucial, I devote Part 3 of this book to it.

Scientists never stop exploring, of course, so even with a well-accepted, well-tested explanation of how the surface of the earth constantly transforms, they don't stop asking questions.

Forging Ahead into New Frontiers

After geologists had the theory of plate tectonics laid out, they had a framework within which they could propose and test specific hypotheses to fill in the details. This work continues today, right now, as you read these words! The frontiers of earth science are being expanded in many directions. In this section, I describe just a few areas of current, exciting research and discovery.

Asking how, where, and why: Mountain building and plate boundaries

Plate tectonics theory explains that the movement of plates creates mountains by pushing crustal rocks together and up (see Chapters 9 and 10). But scientists have not gathered enough evidence to agree on what forces drive the uplift of mountains. Some suggest that a pushing force, exerted by the neighboring plate, forces the rocks upward. Others suggest that the removal of rocks by erosion (explained in Part 4) leads the continental rocks to "float" upward, like an iceberg melting in the ocean.

In Chapter 8, I present a line drawing of plate boundaries. It may seem very straightforward, with lines neatly separating continental plates from one another. But some areas of that map are almost unknown, and the lines have been drawn based on best-guess estimates. In regions such as the northeast Pacific plate, near Kamchatka (a peninsula in eastern Russia), researchers today map earthquake and volcano events in an attempt to pinpoint plate boundaries.

Mysteries of the past: Snowball earth, first life, and mass extinctions

Later in this book (in Chapter 16), I explain exactly how long and complex Earth's history is. Many of the events in Earth's history can be interpreted from patterns in rocks that scientists observe today. The downside to a history told in rocks is that many chapters are missing. These gaps in the record of Earth's past provide fascinating topics for further scientific exploration.

Snowball earth

A hypothesis currently being debated proposes that at some point (between approximately 600 million and 1 billion years ago) the entire planet was covered with ice. This idea is called the *snowball earth hypothesis.*

Some of the evidence to support this hypothesis includes rock formations that are the result of massive layers of ice (glaciers, which I describe in Chapter 13) covering the continents near the equator (at that time). Some scientists argue that an earth covered in ice could not sustain life, and there is evidence of life in rocks from both before and after the suggested time of the snowball. Others want to know what caused the snow and ice to eventually melt. Another hypothesis based on the same evidence suggests that rather than a snowball, the earth was merely a "slushball" that could have, in some areas, still supported life during an extended, very cold period.

Will the snowball earth hypothesis fade into history as a fanciful idea? Or will it be revived, perhaps proven partially true by future studies and incorporated into an accepted geologic theory? Only time (and more research) will tell.

Earliest life

Fossils found in rocks provide a long history of life on Earth, going back nearly 3.6 billion years. These early life forms were tiny, single-celled, simple organisms, such as bacteria. But even at that level, life is a very complicated thing. How did nonliving matter become living matter? Scientists suspect that energy of some kind acted on chemical elements, creating the proper combination to spark life.

They have even re-created such a scenario in a laboratory. But until fossil evidence in the rocks is found that can provide clues about the nature of earliest life, the question is still up for debate.

Mass extinctions

Long after the first life forms existed, Earth experienced periods when many different species thrived, filling the oceans and eventually the land. At least five times in Earth's long history, thousands of species were wiped out in a very short time. (Geologically speaking, a "short time" can span a few million years; I explain geologic time in Chapter 16.) Such events are called *mass extinctions*.

Even if you haven't heard about the other extinctions, you likely know about the extinction of the dinosaurs. But the extinction of the dinosaurs was not the largest mass extinction event in Earth's history. Hundreds of millions of years before the dinosaurs, an extinction took place that killed 80 percent of all the plant and animal groups existing at the time.

In Chapter 22, I describe what is currently known about the mass extinctions in Earth's past. However, geologists and *paleontologists* (people who study fossils) have many unanswered questions about how and why these periods of major extinction occurred. Some propose changes in climate as the culprit, and others point to meteor impacts or extreme volcanic activity. Still others claim only a combination of all these factors could have led to such dramatic mass extinctions.

Predicting the future: Earthquakes and climate change

Scientists in many different fields hope someday to understand enough about Earth's systems to be able to predict what changes may occur in the near future. Two examples I describe here are efforts to predict earthquakes before they occur and the science of future climate change.

Earthquake warnings

You may have firsthand experience with the literally earth-shaking event of an earthquake. If not, you certainly have seen news reports of the terrible devastation that occurs in some regions of the world when strong earthquakes occur. The ability to predict an earthquake event could lead to lifesaving preparations such as evacuation. Much research is focused on looking for early warning signs of an impending quake, with the hope that we could use early warning systems to initiate evacuations and reduce the damage to human lives that occurs with such events.

Researchers have had little luck successfully predicting most destructive earthquakes. Occasionally a large earthquake will be preceded by smaller earthquakes, slight tremors, volcanic activity, or changes in land level relative to sea level. Such was the case in Haicheng, China in 1975, when warnings a day before a large earthquake occurred saved many lives. However, very few earthquakes send advance warning signals.

Current research around the Pacific Ocean focuses on trying to measure the amount of strain being put on two crustal plates as they press against one another. (When the pressure builds to a certain point and is released, an earthquake occurs; see Chapter 10 for details.) Unfortunately, many complex factors lead to an earthquake event, which makes the effort to predict them very challenging. Fortunately, scientists love a good challenge!

Climate change

In looking into Earth's past, one area of intense study is *paleoclimatology*, the study of past climates. Scientists called *paleoclimatologists* take long, cylinder-shaped samples called *cores* from ice sheets. In these ice cores, they find trapped gases and dust from the ancient atmosphere that provide clues to the earth's temperatures long ago. Similar cores of sediments in the bottom of lakes or the ocean may have fossil remains of microscopic organisms. These remains of plant and animal life help scientists called *paleoecologists* build a picture of the ancient environment and past climates.

These are just two of the types of records scientists use to understand Earth's past climate conditions. By combining multiple records and including different types of data, paleoclimatologists and paleoecologists build a picture of climate change throughout Earth's history.

Through these studies, scientists have learned that the earth's climate has gone through dramatic shifts of warming and cooling in the past. Factors including Earth's orbital characteristics and the position of the continents are thought to have affected the climate.

By building a more complete understanding of what changes occurred in the past, scientists hope to be able to predict what changes may occur in the future, particularly in light of industrial civilization's measurable impacts.

Out of this world: Planetary geology and the search for life

Geology is no longer confined to Earth. Advances in scientific understanding of the earth's planetary systems have helped scientists apply that understanding to other planets. Fields of current research include the search for extraterrestrial, or alien, life.

When *astrogeologists* — planetary geologists — look at the surface of Mars, they see features that remind them of features on Earth created by running water. While there is no water on Mars's surface now, these features suggest that large amounts of water once flowed across the surface, presumably originating from underground sources on the planet.

The exploratory Mars rover project is currently collecting sediment samples and other evidence that suggest there may be water not far below the surface of Mars. New data is examined from the rovers all the time and adds to scientists' understanding of Mars's planetary processes.

Why does it matter if there is water on Mars (or any other planet)? One reason is that life on Earth requires water, which means if there is water on another planet, there may be life. Another reason scientists are interested in extraterrestrial water is that water will be necessary for any future human settlements on Mars. It may sound like a movie plot, but it's real-life science!

Chapter **4**

Home Sweet Home: Planet Earth

Y ou could describe the earth as a ball of rock spinning through space, but the earth is made of much more than just rock. Planet Earth is multilayered. If you could travel from the moon to the earth's core, you'd pass through layers that contain gas, liquid, rock, and metal. In this chapter, I briefly describe the various layers of the earth, including what scientists know about Earth's interior layers (and how they learned about those layers).

Earth's Spheres

The materials of Earth's planetary system can be separated into *spheres*, or parts. These five major spheres of Earth's system are illustrated in Figure 4-1:

» **Atmosphere:** The *atmosphere* is a layer of gas that surrounds the entire planet. It serves the important role of protecting everything on Earth from being destroyed by the heat and radiation from the sun and makes life possible on Earth. Within the atmosphere, gases interact with water, forming *weather systems* that circulate air and clouds around the globe.

>> **Hydrosphere:** The *hydrosphere* includes all the water on Earth. The *hydrologic cycle* is the rotation of water through the hydrosphere: flowing as liquid (streams and rivers), evaporating into the atmosphere as gas (clouds), and falling to the surface as rain or snow.

>> **Cryosphere:** The *cryosphere* is composed of all the solid water, or ice, found on Earth's surface. While closely tied to the hydrologic system, the cryosphere can be examined separately because of how massive amounts of surface ice affect the weather and climate systems.

>> **Biosphere:** All the organic materials on Earth — both living and dead organisms — are part of the *biosphere.*

>> **Geosphere:** The solid, rocky layers of the Earth, from the outermost crust to the very center, compose the planet's *geosphere.* Within the geosphere, scientists have further divided the layers of rock material, which I describe in the next section of this chapter.

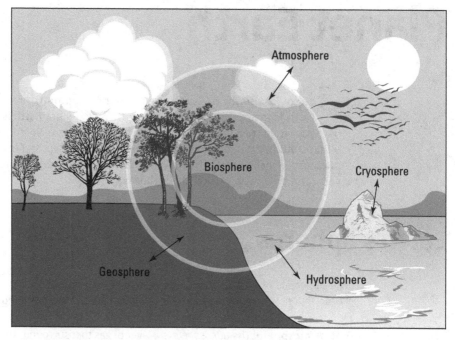

FIGURE 4-1:
The five major spheres of Earth's planetary system.

The earth's spheres are connected to one another through a series of interactions. For example, rainfall from the hydrosphere causes movement of surface materials in the geosphere (a process called *erosion,* which I describe in Part 4). Rainfall also provides water for plants to grow in the biosphere. The interactions among spheres can be studied as subsystems of the earth. An example of a subsystem is the *climate system*, which is influenced by the interaction of atmosphere, biosphere, hydrosphere, and even geosphere.

REMEMBER

Every system needs energy to fuel its processes. Systems on Earth's surface are fueled by heat energy from the sun, whereas other systems (particularly those in the geosphere) are fueled by heat energy that originates deep within the earth.

Because Earth is one giant system, geologists study not only the rock materials on Earth but also how rocks in the geosphere interact with all the other spheres.

Examining Earth's Geosphere

Many geologists study portions of the earth that can be seen. However, some of the most fascinating and still unanswered questions about the earth have to do with what is going on inside — beneath the rocks we can see and touch at the surface.

Humans do not yet have technology advanced enough to dig more than about 12 kilometers (about 7.5 miles) into the earth's crust. So how do scientists know anything about the inside of the earth? They combine their observations of rocks on the surface with knowledge gained from laboratory experiments of temperature and pressure on different materials. Doing so gives them a pretty solid basis on which to make inferences about what occurs in places that can't be directly observed.

Defining Earth's layers

One way scientists separate the layers of Earth's geosphere is by *physical properties*, or whether the layers are liquid or solid.

Because geologists cannot see inside the earth, they make observations about Earth's internal properties by proxy: by interpreting information from earthquake waves that can be used to make inferences about the physical properties of Earth's interior.

REMEMBER

When earthquakes occur, they send out waves. Two types of seismic waves, called *S waves* and *P waves*, are used by scientists to learn about the interior of the earth. These seismic waves are recorded by instruments called *seismometers*, which are buried underground all over the planet. When an earthquake occurs, the seismometer sends a signal from underground to a machine in a lab (a *seismograph*) that records the earthquake wave movements on a printout called a *seismogram*.

Scientists watch the seismographs as they print the seismograms to see when the P waves and S waves arrive. Here's why:

>> P waves travel quickly through solid materials and slow down, slightly changing direction, as they move through liquid materials. By recording where each P wave starts and how long it takes to reach the other side of the planet, scientists have recognized that it must move through regions of solid and liquid materials within the earth.

>> S waves travel through solid materials but cannot travel through liquid at all. When scientists record the path that S waves take through the earth, they find that some S waves never reach the other side — they simply disappear, suggesting that they have hit a section of liquid material.

Figure 4-2 illustrates how the earthquake waves would travel if the interior of the earth was made of one continuous, solid type of material. And Figure 4-3 illustrates how P waves and S waves actually travel through Earth, illustrating the different physical properties of its interior.

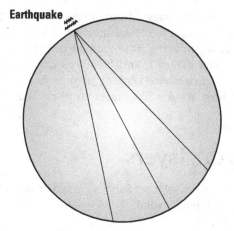

Earthquake

FIGURE 4-2:
The path of wave travel if Earth's interior were a continuous solid.

The areas on the other side of the globe where P waves or S waves do not appear because they either disappear or are *refracted* (change direction) are called *shadow zones.* For more details about P waves, S waves, and shadow zones and why geologists study them, be sure to read Chapter 10.

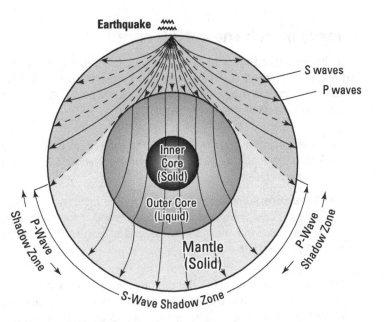

FIGURE 4-3:
The recorded
path of P waves
and S waves.

Examining each layer

Another way scientists categorize the layers of Earth's geosphere is by their composition, or the types of elements and minerals (explained in Chapter 5) that are found in each layer. The different layers of the earth, their estimated depth, and their physical properties are illustrated in Figure 4-4.

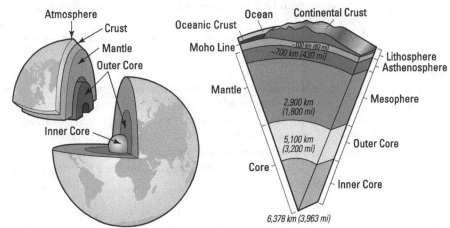

FIGURE 4-4:
The layers of the
earth.

Heavy metal: The earth's core

At the center of the earth is its *core*. Scientists have not been able to sample it directly, but based on their laboratory experiments and interpretations, they believe the core is composed of massive, heavy metal elements such as nickel and iron. The core itself has two layers:

>> **Inner core:** The inner core at the very center of the earth is probably solid and starts at approximately 5,150 kilometers (about 3,200 miles) from the earth's surface.

>> **Outer core:** Surrounding the inner core is a liquid layer of heavy metals called the *outer core*. The study of seismic waves and shadow zones allows scientists to determine that the outer core begins at approximately 2,890 kilometers (1,795 miles) into the earth.

There is no way to measure the temperature of the earth's core. Scientists called *geophysicists* use laboratory studies of iron under conditions of extreme pressure to estimate how hot it may be at such depths. Their estimates range from 5,000 degrees F to 15,000 degrees F. More accurate measurements cannot be made because the conditions of temperature and pressure at the earth's core are much too intense to be re-created in a laboratory setting.

Flowing and solid: The earth's mantle

Outside the earth's metal core is a layer of rock composing the *mantle.* Mantle materials are made of minerals that combine light elements (such as silica and oxygen) with heavier elements (such as iron and magnesium).

Similar to the core below it, the mantle has layers that respond differently to the movement of earthquake waves:

>> **Mesosphere:** In the lower mantle, surrounding the outer core, temperatures are high enough to melt rock, but the intense pressures found so deep in the earth keep mantle materials solid. This deep, solid part of the mantle is called the *mesosphere.* This layer of the mantle begins about 660 kilometers (410 miles) below Earth's surface and continues to where it meets the outer core. Temperatures in the mesosphere range from 3,000 degrees F to almost 8,000 degrees F near the outer core.

REMEMBER

» **Asthenosphere:** The upper part of the mantle is called the *asthenosphere* and exhibits a special physical property. Mantle rocks are solid, but they flow in a way that you may associate with a thick liquid. When solid materials flow in this way, it's called *plastic flow*. I describe plastic flow further in Chapter 13, where I discuss glaciers (which also move in this way). Some geologists describe the asthenosphere as a *crystal mush*, like a cold bowl of oatmeal — mostly solid, but still malleable.

The asthenosphere is found starting at about 200 kilometers (124 miles) from the earth's surface and extends down to the mesosphere. The heat and pressure in this layer can lead to some melting. Due to its ability to flow, this zone of the mantle is considered the weak zone; thus its name *astheno,* which means "weak" or "soft."

COLLECTING MOON ROCKS

The moon is not made of cheese after all. When NASA astronauts visited the moon during the Apollo missions of 1969 to 1972, they collected samples of moon rock material for scientists to test. It turns out the rocks of the moon have a composition similar to the earth's mantle. The moon rocks — even those from deep within the moon — are similar to basalt (see Chapter 7) with almost no heavy metals. (The astronauts collected surface rocks and also drilled core samples.)

This finding suggests that the moon did not form at the same time as the earth or from the same cloud of gas that formed the earth. (I discuss the earth's formation in Chapter 18.) Rather, the composition of the moon rocks suggests that the moon was created by materials from the *surface* of the earth, long after the heavy metal core was covered by mantle and crust materials.

While this idea is still being hotly debated, scientists have been testing the hypothesis that the materials that formed the moon were removed from the earth by a giant impact. Another planet-type body, possibly as large as Mars, may have crashed into Earth, vaporizing some of its crustal and upper mantle materials, leaving them to orbit the earth until they came together forming the sphere now known as the moon.

Wouldn't there be a giant crater as evidence of such an impact? At the time, yes, but this was 4.5 billion years ago, and since then much about the earth and its surface has dramatically changed. The best evidence scientists have is the composition of the moon rocks. Using isotope studies, geologists continue to test the giant-impact hypothesis of moon formation.

>> **Lithosphere:** The uppermost portion of the mantle is attached to the underside of the earth's crust, and together they make up the *lithosphere*. This portion of the mantle is approximately 100 kilometers (62 miles) thick, very rigid, and *brittle* — it breaks and cracks rather than bends or flows. Physically, this part of the mantle is just like the rocks of the crust it is attached to (a brittle solid). But this layer is still mantle because of its mineral composition. In other words, when scientists classify the earth's layers based on their composition, they separate the upper portion of the mantle from the crust because they are composed of different minerals. But when scientists classify the earth's layers based on physical properties, both the uppermost part of the mantle and the earth's crust are part of the brittle lithosphere and different from the flowing solid mantle below.

It's only skin deep: The earth's crust

The boundary between mantle rock and crustal rock in the lithosphere is labeled the Moho discontinuity, named after the scientist who discovered it, Andrija Mohorovičić. This boundary, illustrated in Figure 4-4, is where the composition of rocks changes from the more dense mantle rocks to the much lighter crustal rocks, which are composed primarily of silica.

REMEMBER

The layer of crust covering the earth comes in two types: *continental* and *oceanic*. These two types of crust vary in thickness and are composed of slightly different materials:

>> **Continental crust:** The crust that composes the continents is pretty thick. At its thinnest sections, continental crust is about 12 miles thick; at its thickest sections (where there are mountains), it is up to about 45 miles thick. The rocks that compose the continental crust are primarily granites (see Chapter 7).

>> **Oceanic crust:** This crust, which lies under the earth's oceans, is thin — only about 5 miles thick. This type of crust is composed of dark, dense silicate rocks such as basalt and gabbro (see Chapter 7). Oceanic crust is relatively young, being created even now from the eruption of molten rock along ridges in the sea floor (see Chapter 10).

DRILLING FOR THE MOHO

In 1909, Andrija Mohorovičić, a Croatian seismologist, noticed that earthquake waves increased their speed as they moved through the lower part of the earth's rigid lithosphere. He interpreted (correctly) that this meant the lower portion of Earth's lithosphere is made of a different and slightly denser material than the outer portion. The line where the material in Earth's lithosphere changes from the crustal rock to the mantle rock is named the *Moho line* or the *Mohorovičić Discontinuity*.

For decades scientists have attempted drilling deep into the earth, seeking to reach the Moho line between the mantle rock and crustal rock within the lithosphere. In 2005, a team of scientists with the Integrated Ocean Drilling Program (IODP) came close. The core they drilled near the mid-Atlantic ridge in the Atlantic Ocean reached a depth of 1,416 meters (4,644 feet) into the oceanic crust. But the rocks they recovered appear to be made of crustal rock materials rather than the mantle rocks they were seeking. The researchers concluded that they were close to crossing the Moho discontinuity boundary and plan to attempt drilling a new hole.

Every inch closer to the Moho provides scientists with new information about the composition and formation of the earth's outermost layer and offers clues to the internal structure of Earth's lithosphere.

2
Elements, Minerals, and Rocks

IN THIS PART . . .

Discover the chemistry of elements and compounds.

Learn the basics of minerals.

Take a crash course in rocks.

Chapter **5**

It's Elemental, My Dear: A Very Basic Chemistry of Elements and Compounds

To understand earth processes such as the formation of rocks, it helps to understand some of the basic concepts of chemistry. The science of chemistry explores and describes the properties of substances — gas, liquid, or solid — and explains how and why different substances interact with each other.

In this chapter, I explain that all earth materials are made of atoms, and I show how these atoms interact with one another to create the observable characteristics of rocks and geologic features in the world around you.

The Smallest Matter: Atoms and Atomic Structure

REMEMBER

All matter is made of *atoms*. Every single speck of gas, liquid, or solid surrounding you is a mix of millions of atoms. An atom is the smallest bit of matter that can be measured and identified as a specific element.

Atoms themselves are composed of smaller, *subatomic particles* called neutrons, protons, and electrons. Figure 5-1 is a diagram of atomic structure, including the location of each different subatomic particle type. Protons and neutrons are located in the nucleus at the center of the atom. In each atom, the electrons surround the nucleus, organized into orbital shells. The innermost orbital shell of any atom contains no more than two electrons; the second orbital shell contains no more than eight; and each of the outer shells, while chemically stable with eight electrons, can hold more.

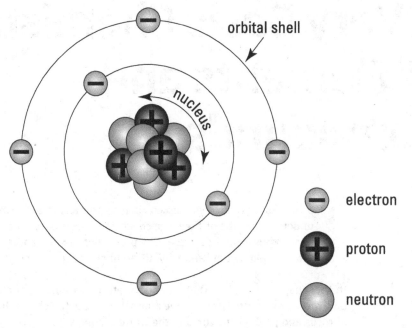

FIGURE 5-1:
The parts of an atom.

REMEMBER

The number of protons in an atom's nucleus determines which element the atom is. For example, an atom with six protons is an atom of the element carbon. An atom with seven protons is an atom of nitrogen.

Getting to know the periodic table

The *periodic table of elements* lists the known elements in order of their *atomic number*, which is their number of protons. Each square on the periodic table provides you with all the information you need to know about that element and how it will interact with other elements. Figure 5-2 illustrates what the different numbers for each element on the periodic table represent:

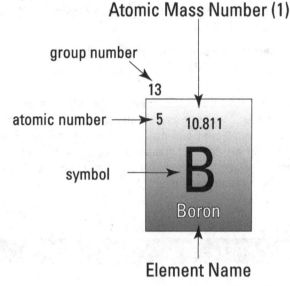

FIGURE 5-2:
The parts of one
square of the
periodic table of
elements.

>> **Atomic mass number:** The *atomic mass number* of an element is the total number of protons and neutrons in its nucleus.

>> **Atomic number:** The *atomic number* of an element is the number of protons in its nucleus.

>> **Group number:** The *group number* tells you how many electrons in the atom are located in the outermost orbital shell and are, therefore, available to bond it to other atoms. For example, elements in Group I have one electron in the outer electron shell, and Group II elements have two electrons in the outer electron shell. The group number for each element may help you understand why some elements, such as Magnesium (Mg) and Calcium (Ca), which are both in Group II, react in similar ways during rock formation and other geologic processes.

TIP

>> **Symbol:** The letters on the periodic table are the symbols for each element. These symbols are a shorthand so that when combinations of elements or chemical reactions are described you don't have to write each element's entire name. The symbols of the periodic table are the same all over the world to make it easier for scientists to communicate.

In many cases, the elemental symbol is based on the name of an element in a different language and may not make sense in your native language. For example, the symbol for gold is Au because in Latin the word for gold is *Aurum,* which means yellow. And the symbol for tungsten is W based on its name in German: *wolfram.*

>> **Element name:** Some periodic tables also list the name of the element below the symbol (see Figure 5-3).

Table 5-1 lists the most common elements in Earth's crust and their approximate percentage. (This list does not represent the proportion of elements in the mantle, nor does it include the iron and nickel that are found in the earth's core, as I describe in Chapter 4.) These elements are the ones that compose nearly all the rocks on Earth's surface. You see them often in this book, so it's a good idea to get familiar with their atomic symbols.

TABLE 5-1

Common Elements in Earth's Crust

Element	Atomic Symbol	% of Crustal Material
Oxygen	O	46.6
Silicon	Si	28
Aluminum	Al	8.1
Iron	Fe	5
Calcium	Ca	3.6
Sodium	Na	2.8
Potassium	K	2.6
Magnesium	Mg	2.1

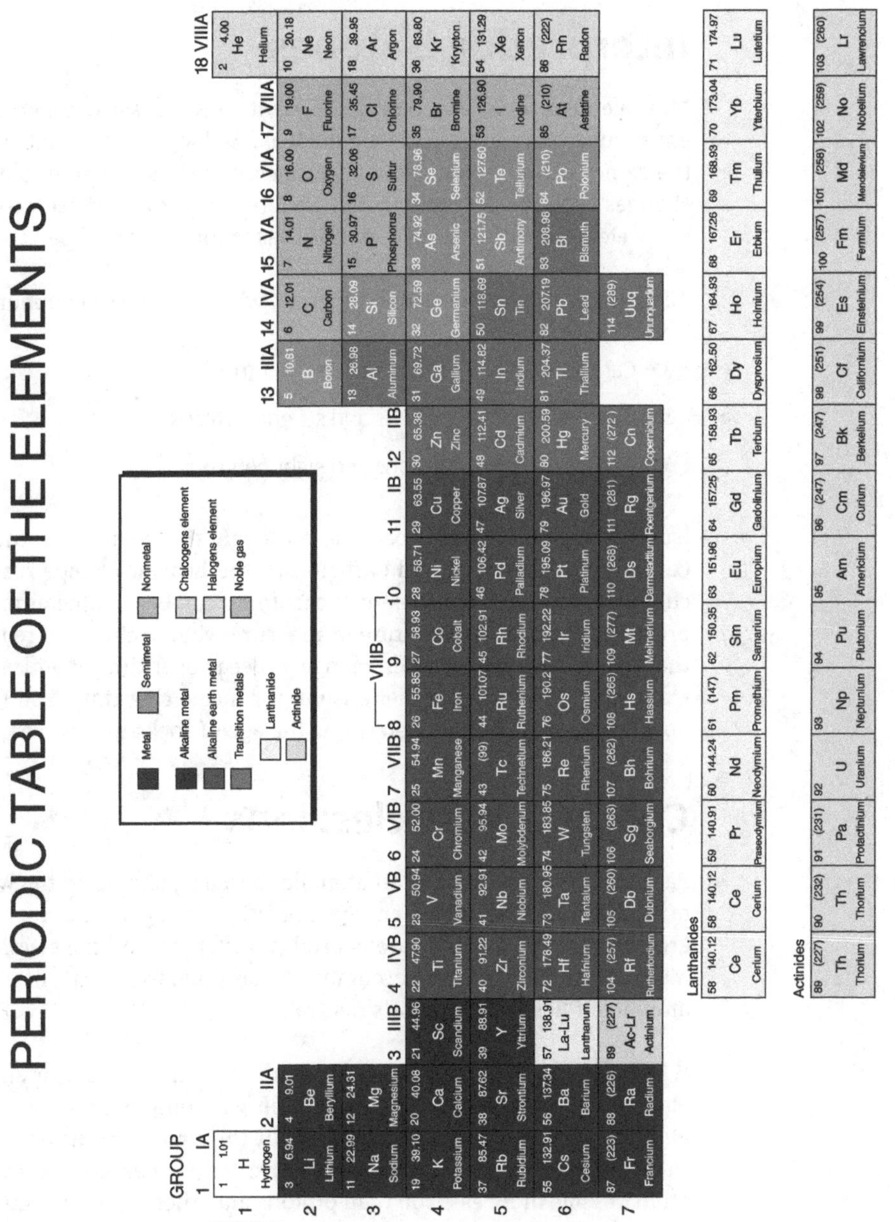

FIGURE 5-3:
The periodic
table of the
elements.

PERIODIC TABLE OF THE ELEMENTS

GROUP

PERIOD

	1 IA	2 IIA		3 IIIB	4 IVB	5 VB	6 VIB	7 VIIB	8 VIIIB	9 VIIIB	10 VIIIB	11 IB	12 IIB	13 IIIA	14 IVA	15 VA	16 VIA	17 VIIA	18 VIIIA
1	1 1.01 H Hydrogen																		2 4.00 He Helium
2	3 6.94 Li Lithium	4 9.01 Be Beryllium												5 10.81 B Boron	6 12.01 C Carbon	7 14.01 N Nitrogen	8 16.00 O Oxygen	9 19.00 F Fluorine	10 20.18 Ne Neon
3	11 22.99 Na Sodium	12 24.31 Mg Magnesium												13 26.98 Al Aluminum	14 28.09 Si Silicon	15 30.97 P Phosphorus	16 32.06 S Sulfur	17 35.45 Cl Chlorine	18 39.95 Ar Argon
4	19 39.10 K Potassium	20 40.06 Ca Calcium		21 44.96 Sc Scandium	22 47.90 Ti Titanium	23 50.94 V Vanadium	24 52.00 Cr Chromium	25 54.94 Mn Manganese	26 55.85 Fe Iron	27 58.93 Co Cobalt	28 58.71 Ni Nickel	29 63.55 Cu Copper	30 65.38 Zn Zinc	31 69.72 Ga Gallium	32 72.59 Ge Germanium	33 74.92 As Arsenic	34 78.96 Se Selenium	35 79.90 Br Bromine	36 83.60 Kr Krypton
5	37 85.47 Rb Rubidium	38 87.62 Sr Strontium		39 88.91 Y Yttrium	40 91.22 Zr Zirconium	41 92.91 Nb Niobium	42 95.94 Mo Molybdenum	43 (99) Tc Technetium	44 101.07 Ru Ruthenium	45 102.91 Rh Rhodium	46 106.42 Pd Palladium	47 107.87 Ag Silver	48 112.41 Cd Cadmium	49 114.82 In Indium	50 118.69 Sn Tin	51 121.75 Sb Antimony	52 127.60 Te Tellurium	53 126.90 I Iodine	54 131.29 Xe Xenon
6	55 132.91 Cs Cesium	56 137.34 Ba Barium	57 138.91 La-Lu Lanthanum	72 178.49 Hf Hafnium	73 180.95 Ta Tantalum	74 183.85 W Tungsten	75 186.21 Re Rhenium	76 190.2 Os Osmium	77 192.22 Ir Iridium	78 195.09 Pt Platinum	79 196.97 Au Gold	80 200.59 Hg Mercury	81 204.37 Tl Thallium	82 207.19 Pb Lead	83 208.98 Bi Bismuth	84 (210) Po Polonium	85 (210) At Astatine	86 (222) Rn Radon
7	87 (223) Fr Francium	88 (226) Ra Radium	89 (227) Ac-Lr Actinium	104 (257) Rf Rutherfordium	105 (260) Db Dubnium	106 (263) Sg Seaborgium	107 (262) Bh Bohrium	108 (265) Hs Hassium	109 (266) Mt Meitnerium	110 (268) Ds Darmstadtium	111 (281) Rg Roentgenium	112 (272) Cn Copernicium	114 (289) Uuq Ununquadium						

Legend:
- Metal
- Semimetal
- Nonmetal
- Alkaline metal
- Alkaline earth metal
- Transition metals
- Chalcogens element
- Halogens element
- Noble gas
- Lanthanide
- Actinide

Lanthanides

58 140.12 Ce Cerium	59 140.91 Pr Praseodymium	60 144.24 Nd Neodymium	61 (147) Pm Promethium	62 150.35 Sm Samarium	63 151.96 Eu Europium	64 157.25 Gd Gadolinium	65 158.93 Tb Terbium	66 162.50 Dy Dysprosium	67 164.93 Ho Holmium	68 167.26 Er Erbium	69 168.93 Tm Thulium	70 173.04 Yb Ytterbium	71 174.97 Lu Lutetium

Actinides

90 (227) Th Thorium	91 (232) Pa Protactinium	92 (231) U Uranium	93 Np Neptunium	94 Pu Plutonium	95 Am Americium	96 (247) Cm Curium	97 (247) Bk Berkelium	98 (251) Cf Californium	99 (254) Es Einsteinium	100 (257) Fm Fermium	101 (256) Md Mendelevium	102 (259) No Nobelium	103 (260) Lr Lawrencium

Interpreting isotopes

Most elements exist as atoms of different atomic mass number, indicating different numbers of neutrons in the nucleus. As long as the number of protons stays the same (the atomic number), you have the same element, but its atomic mass changes with the addition or subtraction of neutrons. These various atoms of the same element with different atomic mass numbers are called *isotopes*.

Take, for example, the element carbon, which has three common isotopes:

>> Carbon-12 has six protons and six neutrons.

>> Carbon-13 has six protons and seven neutrons.

>> Carbon-14 has six protons and eight neutrons.

TIP

Isotopes are very useful because although the element is the same (such as Carbon-12, Carbon-13, and Carbon-14), the heavier isotope reacts differently in chemical reactions. This means the isotopes can be counted or measured to interpret conditions of temperature or pressure when a chemical reaction occurred in the past. Also, some isotopes change or decay over time at a measurable and constant rate, which makes them useful for measuring time. You find details about how isotopes are used to determine the age of rocks in later chapters.

Charging particles: Ions

Each subatomic particle in an atom has a *charge*, similar to the way opposite ends of a battery or magnet are charged: positive or negative. In an atom, the protons are positive, the neutrons are neutral (no charge), and the electrons are negative. Most atoms have the same number of protons and electrons, which means the atom itself has no charge; it's neutral.

When an atom with only one electron in its outer shell is near an atom with seven electrons in its shell, the single electron will jump over to join and complete the almost-full shell. This action results in the first atom having one more proton than electrons and, therefore, a positive (or +1) charge. Meanwhile, the second atom has one more electron than protons and, therefore, a negative (or −1) charge. (Later in the chapter, Figure 5-4 illustrates this fact.)

REMEMBER

Atoms or *molecules* (more than one atom joined together) with positive or negative charge are called *ions*. The charge of the ion is determined by how the electrons in its outer shell move to and from nearby atomic shells. An atom with a positive charge is called a *cation*, and an atom with a negative charge is called an *anion*. Atoms, and even compounds, can have negative charges of 1, 2, 3, and even 4 (though 4 is rare) and positive charges up to +8. The interaction of ions with one another is one way that atoms form bonds; keep reading to find out the details.

Chemically Bonding

Very few atoms exist in nature all by themselves. Multiple atoms joined together are called *molecules*. Some atoms of the same element pair up with each other to form molecules. An example of this is oxygen gas, which is composed of two oxygen atoms, written as O_2. (The small 2 indicates how many atoms of oxygen are in the molecule.)

In other cases, atoms of two or more different elements combine to form a *compound*. The compound is held together by a *chemical bond*. In this section, I explain the three most common types of chemical bonding between atoms.

REMEMBER

How two atoms bond together is determined by the number of electrons in their outer orbital shells. For example, an atom with 13 total electrons such as aluminum (Al) will have two electrons in the first orbital shell, eight electrons in the second orbital shell, and three electrons in the outermost orbital shell. The three electrons in the outermost shell are the ones that participate in bonds with other atoms.

Donating electrons (ionic bonds)

When two atoms trade electrons between their outer orbital shells, becoming a cation and an anion, they form an *ionic bond*. The result of an ionic bond is that the positively charged cation and negatively charged anion combine into a compound that has a neutral charge. All ionic bonds create compounds called *salts*. Of these compounds, you are most familiar with table salt, or NaCl. In this molecule, an atom of sodium (Na) and an atom of chlorine (Cl) have bonded together. They are held together because the single electron in the outer shell of the sodium atom has been donated to fill the outer shell of chlorine (which had only seven electrons to begin with). This bond is illustrated in Figure 5-4.

Sharing electrons (covalent bonds)

When two atoms bond together, and neither one donates or gives up an electron, they form a covalent bond. In a *covalent bond* the atoms share the electrons in their outer orbital shells. The sharing of electrons in covalent bonding creates a very strong bond because each atom participating in the electron share has a full outer shell and a neutral charge.

An example of a covalent bond is found in a molecule of water, H_2O. As illustrated in Figure 5-5, the two atoms of hydrogen (H) share electrons with the atom of oxygen (O).

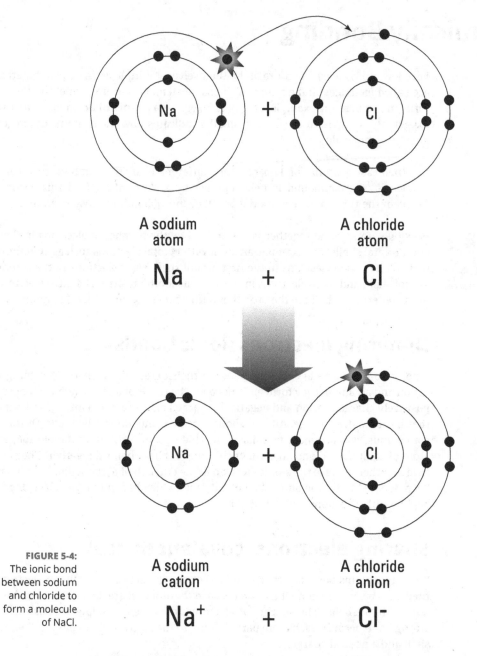

A sodium
atom

A chloride
atom

Na   +   Cl

A sodium
cation

A chloride
anion

Na⁺   +   Cl⁻

Migrating electrons (metallic bonds)

Metallic bonds occur between atoms that have very few electrons in their outer-most electron shells. Instead of donating or sharing these electrons, the electrons are released from the orbital shell and available for a nearby cluster of atoms to use.

Some scientists describe the metallically bonded atoms as floating in a sea of electrons because the electrons are not specifically attached to the orbital shell of any atom in particular. The electrons in a metallic bond move freely from one atom to another. This idea is illustrated in Figure 5-6.

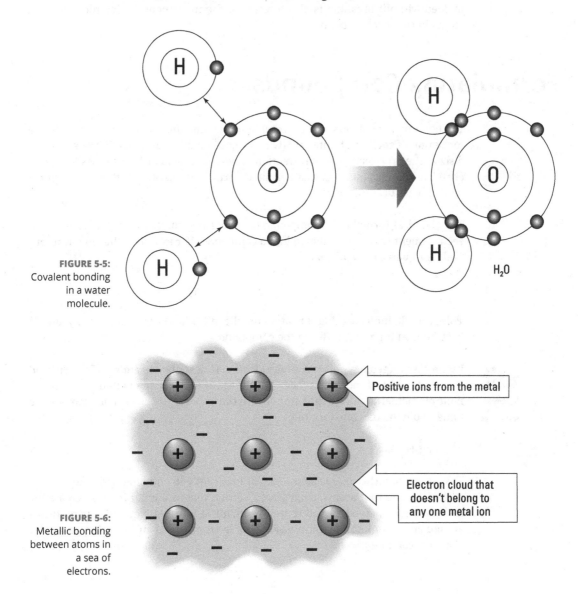

FIGURE 5-5:
Covalent bonding in a water molecule.

H_2O

Positive ions from the metal

Electron cloud that doesn't belong to any one metal ion

FIGURE 5-6:
Metallic bonding between atoms in a sea of electrons.

The unique nature of metallic bonds is what gives metals such as gold or silver their unique characteristics. The ability to conduct electrical current is a result of the movement of electrons. The shiny, or metallic appearance is due to the large number of freely floating electrons. And the fact that metals can be bent and molded without breaking is also a result of the movement of electrons between atoms in the metallic bond.

Formulating Compounds

The bonding of elements to form compounds is fundamental to understanding the formation of rocks and minerals (which I describe in Chapter 6). When scientists discuss the processes of rock formation, as well as other earth processes involving chemical changes (such as weathering, described in Chapter 7), they use a short-hand of chemical formulas.

The chemical formula of a compound describes the number of different atoms of each element that are combined into a compound. For example, the chemical formula for quartz is as follows:

SiO_2

This formula indicates that one atom of silicon (Si) and two atoms of oxygen (O) have bonded together, forming the compound.

In the case of geology, most chemical formulas describe *minerals,* which are solid structures built of molecules (see Chapter 6). In mineral compounds, sometimes multiple elements can fill the same spot in the mineral structure. For example, the mineral olivine has this formula:

$(Mg, Fe)_2SiO_4$

Two atoms of either magnesium (Mg) *or* iron (Fe) will combine with one atom of silicon (Si) and four atoms of oxygen (O). Either magnesium or iron can create the mineral olivine, so when you write the chemical formula, you put a parenthesis around and a comma separating the possible atoms that can form that particular chemical compound.

Chapter **6**

Minerals: The Building Blocks of Rocks

N early every rock on the earth is built of minerals. You are probably familiar with gemstone minerals found in jewelry, but the less eye-catching minerals are what make up most rocks.

Every mineral is a combination of elements; the atoms are organized into geometric structures called *crystals*. In many rocks the mineral crystals are too small for you to see without a magnifying glass or microscope, but they're still there.

In this chapter, I show you how elements are organized into crystals and how minerals are identified. I also give you an introduction to silicate minerals (the types of minerals found in most rocks in the earth's crust) and the less-common (but still important) nonsilicate minerals.

Meeting Mineral Requirements

To be considered a mineral, an earth material must meet these requirements:

>> **It's solid.** Minerals are solid — not liquid or gas — at the temperature of Earth's surface.

>> **It's inorganic.** Minerals are *inorganic,* meaning they are not built of the carbon-based organic compounds that make up living tissue in plants and animals. However, many animals build mineral skeletons or shells. These minerals, such as the apatite in your skeleton, or the calcium carbonate in a seashell are biogenic, or created by a lifeform, but not, in themselves, living.

>> **It has an orderly structure.** When atoms combine to form minerals, they do so in an organized way that forms a geometric pattern called a *crystal.* Earth materials that form without this orderly crystalline structure are described as *amorphous* or *glassy* and are not minerals, though they do form certain rocks, like chert and obsidian (see Chapter 7 for more on these amorphous and glassy solids).

>> **It is naturally occurring.** True minerals form in nature and are built through natural processes of chemical bonding. Humans have the technology to combine atoms into crystals in a lab and mimic the processes that occur in nature, but true minerals occur naturally.

>> **It has a specific chemical composition.** Each mineral has a defined *chemical composition:* a combination of elements that creates its particular crystal structure. Some minerals may have multiple, similar chemical compositions, where similar-sized atoms of different elements may replace each other. In these cases the small range of variation is well-defined for each mineral.

Making Crystals

Minerals come in many shapes. The shape of a mineral is determined by how the atoms creating the mineral are organized. This three-dimensional shape formed by the bonded atoms is the *crystal structure* (sometimes called the crystal *lattice*) of the mineral. When the atoms bond together, they arrange themselves in a partic- ular pattern.

REMEMBER

No matter how large or small, all the crystals of a particular mineral have the same crystal structure. Minerals with more than one common crystal shape are said to have multiple *habits.* For example, the mineral pyrite (called *fool's gold*) has more than one commonly occurring habit, or crystal shape: Sometimes pyrite minerals form perfect cubes (with six sides), and other times they form octohedrons, with eight sides.

You may have seen large crystals such as the ones shown in this book's photo insert, and you've undoubtedly seen gemstones in jewelry, which are large crystals as well. Large crystals are minerals, but the minerals in rocks are often much smaller. Here are two examples of why you find smaller minerals in many rocks:

>> The minerals in rocks may have been broken down into small pieces. This is the case in most sedimentary rock, which I explain in Chapter 7.

>> The rock minerals may have formed under conditions that prevented them from growing large. An example of this situation are the minerals that create a rock called *basalt*, which I also describe in Chapter 7.

REMEMBER

Which minerals make up a rock depend on which elements are present, as well as the conditions of temperature and pressure where the rock is forming. For example, diamond and graphite are different minerals formed by the same element: carbon. The crystal structure of graphite (commonly used as pencil lead) is formed under conditions of low temperature and low pressure, similar to conditions on Earth's surface. Diamond crystals, while also composed of only carbon, are formed under conditions of very high temperatures and pressures, deep within the earth.

Identifying Minerals Using Physical Characteristics

Because each mineral is composed of certain elements arranged in a specific crystal structure, physical characteristics can be used to identify it. In this section, I describe the most common physical characteristics, or *properties,* used to identify minerals.

Observing transparency, color, luster, and streak

Some of the most obvious observations you can make about a mineral relate to its appearance in relation to light:

>> **Transparency:** One visible characteristic of a mineral is its ability to *transmit* light, or allow light to pass through it. This property is sometimes called its *clarity*. A mineral that you can see through is *transparent*. A mineral that allows

light to pass through but is not clear enough to see through is *translucent*. And a mineral that no light passes through is *opaque*.

>> **Color:** Color is a result of how light is absorbed or reflected by an object. In the case of minerals, their color can be changed in many ways that do not affect the crystal structure or composition. For example, heating a mineral may shift atoms and result in color change, or very small amounts of other elements, called *impurities*, may be trapped in the crystal structure and give it a color.

TIP

For these reasons, color is the least reliable method of identifying minerals. After all, a single mineral may come in many different colors. (This is the secret to many semi-precious gemstones, which I describe later in this chapter and in Chapter 23.)

>> **Luster:** A mineral's *luster* is a description of how the surface reflects light. Luster is either metallic (shiny like a metal) or nonmetallic. Nonmetallic lusters include pearly, glassy, silky (like silk fabric), satin, earthy (dull), greasy, or *adamantine* (extremely shiny, or fiery like a diamond).

TECHNICAL
STUFF

A metallic luster is the result of light transmitting energy to electrons of the surface atoms. This energy causes the electrons to vibrate and emit a shininess you observe as metallic. This type of luster is named after metals, which have a high number of free electrons that respond the same way to light.

>> **Streak:** *Streak* is how the mineral appears in powdered form. To test the streak of a mineral, you rub the mineral on a *streak plate,* which is a piece of rough porcelain. (Streak plates are often included in rock and mineral kits for this particular purpose.) Rubbing the mineral against a streak plate grinds off some of the mineral into a powdered line, or streak, across the plate. Many minerals can be identified by the color of their streak, which may be very different from the color of the mineral sample you hold in your hand. For example, the mineral pyrite appears gold but leaves a black streak when rubbed across a streak plate.

Measuring mineral strength

Other physical characteristics useful for identifying minerals indicate the strength of the crystal structure. These include hardness, tenacity, cleavage, and fracture.

Hardness

Mineral hardness is the most useful way of identifying minerals. *Hardness* is how well a mineral resists being scratched. When you scratch the surface of a mineral you are breaking the bonds between atoms, so hardness is a way of measuring how strong those bonds are.

Mineral hardness is given a value on *Mohs scale of relative hardness*, which is illustrated in Figure 6-1. Mohs scale, named after the mineralogist Friedrich Mohs who developed it, is a *relative scale*, which means it ranks the minerals in relation to one another — not along an absolute measure of hardness.

Mohs Mineral Hardness Scale

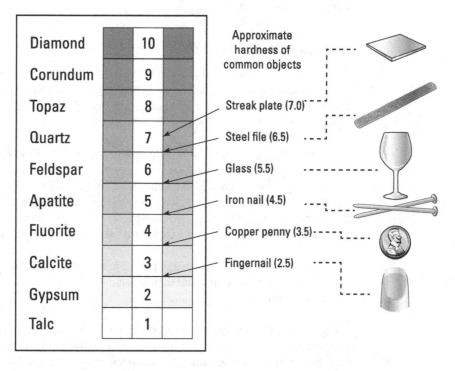

FIGURE 6-1:
Mohs relative
mineral hardness
scale.

TIP

To test for hardness of a mineral, you scratch its surface with other minerals (or objects of known hardness, such as a penny, your fingernail, or a pocketknife) until you narrow down its hardness on the scale. For example, if your fingernail will not scratch a certain mineral but a penny will, your mineral is probably calcite, or another mineral with similar hardness. Diamonds are at the top of the scale because only another diamond can scratch a diamond's surface.

Scientists can also measure a mineral's *absolute hardness* in a laboratory. The values of absolute hardness are very different from the relative hardness. For example, while diamond and corundum are only one point different on Mohs relative hardness scale, in absolute hardness the diamond is more than four times as hard as corundum.

Tenacity

A less commonly used characteristic of mineral strength is its *tenacity* or toughness. Terms describing mineral tenacity indicate how the mineral resists breaking. For example, metallic minerals can be hammered and shaped without breaking into pieces, so they are called *malleable.* Some minerals like mica are *elastic:* They bend and then bounce back to their original form. Many minerals (including pyrite, hornblende, and olivine) are *brittle*, which means they break fairly easily into smaller pieces.

Cleavage and fracture

When you take your hammer and tap (or smash) a mineral, the way it breaks tells you important information about its crystal structure and molecular bonds. The way a mineral breaks can be categorized in one of two categories: cleavage or fracture.

If a mineral has layers of weaker and stronger bonds, it will *cleave* (break) along *cleavage planes:* planes of weakness in the crystal structure. These cleavage planes produce flat surfaces and angled geometry that are useful for identifying the mineral. The bonds between atoms along the cleavage surfaces are weaker than other atomic bonds in the minerals. Two minerals that are the same in other characteristics may have different cleavage planes that allow you to accurately identify them.

A few examples of mineral cleavage planes are illustrated in Figure 6-2. Mineral cleavage is described as the number of planes (flat surfaces) and the angle at which they intersect.

In Figure 6-2, the sketch of muscovite has one cleavage plane and no angles of intersection, forming sheets. Feldspar has two planes of cleavage that intersect at a 90-degree angle (a right angle). Calcite has three cleavage planes, but they do not intersect at right angles. And halite has three planes of cleavage intersecting at 90-degree angles, which create cube shapes. There are also minerals with four cleavage planes (like fluorite) and even five or six (such as the mineral sphelurite).

If all the bonds within the crystal are equally strong, the mineral will *fracture* instead of cleave. A mineral that fractures often fractures *irregularly*, or with rough, uneven surfaces.

Glassy or amorphous solids such as obsidian and chert do not have an underlying crystal structure like minerals do. These solids fracture in a special way due to their lack of crystal structure. Breaking a piece of obsidian or chert leaves a smooth, curved surface, similar to a clamshell in appearance. This type of fracture, called a *conchoidal* fracture, is illustrated in Figure 6-3.

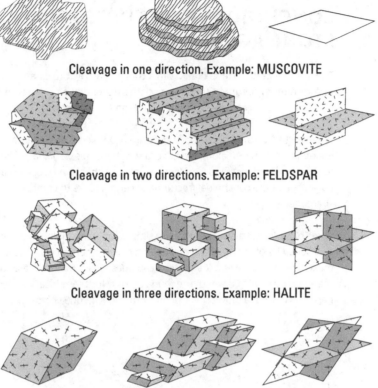

Cleavage in one direction. Example: MUSCOVITE

Cleavage in two directions. Example: FELDSPAR

Cleavage in three directions. Example: HALITE

Cleavage in four directions. Example: FLUORITE

FIGURE 6-2:
Cleavage planes of muscovite, feldspar, halite, and flourite.

FIGURE 6-3:
The appearance of a conchoidal mineral fracture.

CONCHOIDAL FRACTURES AND STONE AGE TOOLS

The fact that certain rocks break with a conchoidal fracture came in very handy for prehistoric human cultures. Stone Age societies depended on rocks with conchoidal fracture characteristics to create sharp-edged tools and weapons.

When rocks fracture conchoidally, they break off pieces of stone, called *flakes*. With enough flakes broken along the edge, a rock becomes very sharp, like a knife. Stone toolmakers used techniques of *flint-knapping*, or specialized rock-breaking, that took advantage of the conchoidal fracturing of certain rocks to create useful shapes and sharpened edges.

Some of the earliest stone tools were simple chopping tools, where only one edge was sharpened by breaking off pieces. But later, as the human brain developed and toolmakers refined their flint-knapping techniques, beautifully intricate stone tools, such as the spear points of Clovis culture found across North America, were created. The figure here shows what a Clovis spear point from about 13,000 years ago looks like.

If it tastes like salt, it must be halite: Noting unique mineral properties

Some minerals have distinguishing properties of taste or smell. If you have ever seen a geologist licking a rock, this may be the reason. For example, the mineral halite is salty. Because halite looks just like calcite, the salty taste is a useful way to distinguish halite from calcite.

WARNING

Aside from halite, licking is not a useful way of identifying minerals. In fact, licking unknown minerals may expose you to poisonous metals such as arsenic and lead and is not recommended.

Another unique mineral property is the *effervescence,* or fizzing, of certain minerals when sprayed with weak hydrochloric acid. This property is unique to minerals made of calcium carbonate ($CaCO_3$), because the acid (HCl) reacts to break the ionic bond in the mineral and produces a new ionic bond (CaCl) as well as water (H_2O) and carbon dioxide (CO_2) gas, which is what you see escaping in the bubbles! If you do not have a taste for rocks, this method is another way to distinguish calcite from halite: If you spray diluted hydrochloric acid on the mineral and it fizzes and bubbles, you have calcite; if nothing happens, you have halite. (After you have put acid on the mineral, you definitely don't want to lick it to confirm it is halite!)

Measuring properties in the lab

Some mineral properties can be measured only in a laboratory. To examine minerals under a microscope, geologists make *thin-section* slides, which are very thin and polished slices of the rock they want to examine.

A thin-section slide is made by attaching a small piece of the rock to a glass slide and grinding down the rock until it is thin enough for light to shine through it. Then the thin-section can be examined under a special geologic microscope, called a *petrographic* microscope, that shines light through the slide. The petrographic microscope has settings that change the angle of the light shining through the minerals — a process called *polarization.*

When light passes through a crystal, it is separated and different colors appear depending on the crystal shape and composition. The *polarized light* highlights the internal structure of the crystal, which can indicate the types of minerals in the rock.

Other mineral properties measured in a lab include:

>> **Fluorescence:** *Fluorescent* minerals in a rock glow when you shine ultraviolet light on them.

>> **X-ray diffraction:** Each mineral displays a particular pattern when x-rays pass through it and are separated, or *diffracted.* The distinct *x-ray diffraction* patterns indicate which minerals are in a rock.

Realizing Most Rocks Are Built from Silicate Minerals

The minerals commonly found in rocks at the earth's surface are collectively called *rock-forming minerals.* Of all the known elements, only eight of them make up most of the rock-forming minerals; these eight elements compose 98 percent of the rocks in the earth's crust. These eight are:

>> Oxygen (O)

>> Silicon (Si)

>> Aluminum (Al)

>> Iron (Fe)

>> Calcium (Ca)

>> Sodium (Na)

>> Potassium (K)

>> Magnesium (Mg)

REMEMBER

Silicon and oxygen are the most common elements in the earth's crust. (Refer back to the pie chart in Figure 2-1, which illustrates the percentage of elements in Earth's crust.) Atoms of these two elements combine to form the basis for a group of minerals called *silicates.* Every silicate mineral begins with a silicon-oxygen *tetrahedron* of atoms like the one illustrated in Figure 6-4. The tetrahedron has four oxygen atoms attached to a central silicon atom creating a pyramid shape.

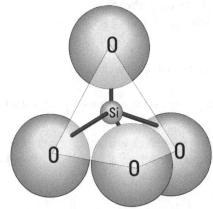

FIGURE 6-4:
The silicon-oxygen tetrahedron.

Multiple tetrahedra combine to form *silicate structures*, which are then held together with atoms of other elements, such as iron, magnesium, potassium, sodium, and calcium.

REMEMBER

The chemical composition and structure of a silicate mineral provide clues to the temperature and pressure conditions under which it was formed. Many silicate minerals form from the cooling of molten rock (magma, described in Chapter 7). Others form under the pressure of mountain building (see Chapter 9) or as surface materials are weathered (see Chapter 7).

Finding silicates in many shapes

The different silicate structures are created by the way multiple silicon-oxygen tetrahedra combine to share oxygen atoms. Five common silicate structures exist:

>> **Single tetrahedral:** Individual tetrahedra that are held together by other elements without sharing any oxygen atoms have the *single tetrahedral* structure, as illustrated in Figure 6-4. The mineral olivine has single tetrahedral structure.

>> **Single chains:** In *single chain* silicates, the tetrahedra share an oxygen atom with two other tetrahedra, forming a linked chain, as illustrated in Figure 6-5. The mineral augite has a single chain silicate structure.

FIGURE 6-5:
The single chain
silicate structure.

>> **Double chains:** *Double chain* structures are created when two single chain structures share a few oxygen atoms, linking the two chains together, as illustrated in Figure 6-6. An example of a mineral with double chain structure is hornblende.

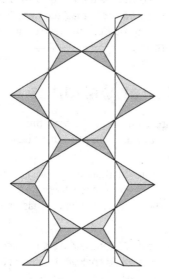

FIGURE 6-6:
The double chain
silicate structure.

>> **Sheets:** A *sheet* silicate structure is formed when the three atoms at the base of the tetrahedra are shared with other tetrahedra, linking them together in a sheet, as illustrated in Figure 6-7. The remaining oxygen atoms of each tetrahedral bond one silicate sheet to another. Biotite is a mineral with sheet silicate structure.

>> **Framework:** A *framework* silicate structure is three-dimensional. Each oxygen atom of the tetrahedral is shared with another tetrahedral, as illustrated in Figure 6-8. An example of a mineral with framework silicate structure is albite.

>> **Ring:** When tetrahedrals share two oxygen atoms in groups of three, four, five, or six, they may be arranged in a circle or *ring* structure, as illustrated in Figure 6-9. The mineral kaolinite has ring structure.

The structure of silicate minerals determines the minerals' cleavage characteristic. That's because the mineral cleaves along its weakest planes. In this case, the bonds linking two structures are weaker than the bonds that hold each structure together (the shared oxygen). Sheet silicate minerals have a single plane of cleavage when broken because the bonds forming the sheet structure are stronger than the bonds between sheets.

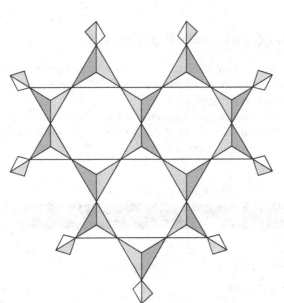

FIGURE 6-7:
The sheet silicate
structure.

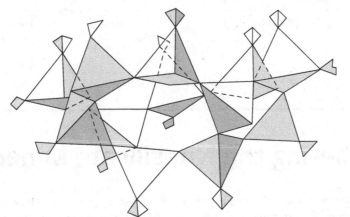

FIGURE 6-8:
The framework
silicate structure.

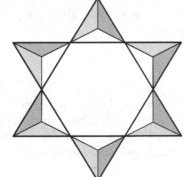

FIGURE 6-9:
The ring silicate
structure.

Grouping silicate minerals

To organize the thousands of silicate minerals, geologists categorize them into groups according to their silicate structure. The seven most common silicate groups and their associated structures are listed in Table 6-1. Remember that when two or more elements in a chemical formula are shown inside parentheses, such as (Mg, Fe) it means either of those elements can fill that spot in the chemical formula and mineral structure.

TABLE 6-1   ## Silicate Mineral Groups

Group Name	Chemical Formula	Example Mineral	Tetrahedra Structure
Olivine group	$(Mg, Fe)_2SiO_4$	Olivine	Single tetrahedra
Pyroxene group	$(Mg, Fe)_2SiO_3$	Augite	Single chains
Amphibole group	$Ca_2(Fe, Mg)_5Si_8O_{22}(OH)_2$	Hornblende	Double chains
Mica group	$K(Mg, Fe)_3AlSi_3O_{10}(OH)_2$ / $KAl_2(AlSi_3O_{10})(OH)_2$	Biotite/Muscovite	Sheets
Feldspar group	$KAlSi_3O_4$ / $(Ca, Na)AlSi_3O_8$	Orthoclase/ Plagioclase	Framework
Quartz	SiO_2	Quartz	Framework
Kaolinite group (clays)	$Al_2Si_2O_5(OH)_4$	Kaolinite	Ring

Remembering the Nonsilicate Minerals

While most rock-forming minerals in Earth's crust are silicates, the *nonsilicate* minerals are important too. Nonsilicate minerals are crystals composed of elements other than silicon. Composing only about 5 to 8 percent of crustal materials, many of the nonsilicate minerals are found in sedimentary rocks (see Chapter 7) and others are economically valuable. Here, I describe a few of the major groups of nonsilicate minerals.

Carbonates

Nonsilicate minerals called *carbonates* are important minerals in sedimentary rocks. The most common carbonate mineral is *calcite*, which is the primary mineral in the sedimentary rocks limestone and dolostone and in the metamorphic rock marble. Table 6-2 lists some common carbonate minerals.

TABLE 6-2

Common Carbonate Minerals

Mineral Name	Chemical Formula	Common Use
Calcite	$CaCO_3$	Cement
Dolomite	$CaMg(CO_3)_2$	Cement

Sulfides and sulfates

Nonsilicate minerals made with sulfur include *sulfides*, such as galena, which provide useful metals, and *sulfates*, such as gypsum, which are used in plaster building materials. These are listed in Table 6-3.

TABLE 6-3

Sulfide and Sulfate Minerals

Mineral Name	Chemical Formula	Common Use
Sulfates		
Gypsum	$CaSO_4 + 2H_2O$	Plaster
Anhydrite	$CaSO_4$	Plaster
Sulfides		
Galena	PbS	Lead ore
Pyrite	FeS_2	Sulfur ore
Cinnabar	HgS	Mercury ore
Chalcopyrite	$CuFeS_2$	Copper ore

Oxides

Oxide minerals form when oxygen bonds with metals. They are the source of some of Earth's most important metals, including iron, aluminum, titanium, and uranium. Common oxides are listed in Table 6-4.

TABLE 6-4

Common Oxide Minerals

Mineral Name	Chemical Composition	Common Use
Hematite, Magnetite	Fe_2O_3, Fe_3O_4	Iron ore
Corundum	Al_2O_3	Gemstone
Cassiterite	SnO_2	Tin ore
Rutile	TiO_2	Titanium ore
Uraninite	UO_2	Uranium ore

Native elements

Native element minerals form when atoms of an element bond to each other, creating a "pure" mineral with no other ingredients. Native element minerals include gold, silver, diamond, copper, and platinum and are listed in Table 6-5.

TABLE 6-5

Common Native Element Minerals

Mineral Name	Chemical Composition	Common Use
Gold	Au	Jewelry, coins, electronics
Silver	Ag	Jewelry, coins, photography
Platinum	Pt	Jewelry, gasoline production
Diamond	C	Jewelry, drill bits
Copper	Cu	Electrical wiring

Evaporites

Another group of minerals, called *evaporites*, are classified according to how they form rather than by their composition. Evaporite minerals form when water they are dissolved into evaporates, leaving the mineral crystals behind. Common evaporite minerals include gypsum and anhydrite, which are listed in Table 6-3 under their compositional category: sulfates.

Other common evaporite minerals include halite (sodium chloride, or table salt) and sylvite (potassium chloride).

Gemstones

Some minerals are considered especially valuable and beautiful and are called *gemstones.* Some of these minerals are valuable because they are rare, such as diamonds or emeralds. But others are very common, such as amethyst or opal. Minerals that are considered gemstones form under conditions that allow the individual crystals to grow fairly large. These conditions occur when molten rock cools very slowly underground, or when rocks are metamorphosed by heat and pressure (see Chapter 7 on metamorphism).

Many of the gemstones you may be familiar with are quartz minerals that have small amounts of other elements, or *impurities*, in their crystals. These impurities change the color of the crystal. For example, amethyst is a quartz crystal with lavender or purple color due to small amounts of manganese in it. Citrine is also a quartz crystal, with small amounts of iron giving it a yellow to orange color.

DIAMONDS

Diamonds form under conditions of high pressure found in the earth's mantle, about 100 miles below the crust. They are brought to the surface by explosive eruptions of molten materials up through the crust. These explosions of molten material cool and form a structure called a *kimberlite pipe* (first discovered in Kimberley, South Africa). Most of these kimberlite pipes seem to have formed during the breaking apart of the supercontinent Pangaea (see Chapter 19). Some have been found in the very oldest rocks of the continents (called the *craton*), and geologists think they were created back when the earth's interior was much hotter, producing more explosive eruptions on the early continents. Still, only some of these locations produce diamonds large enough to be gem quality, making them rare and valuable.

Scientists have created tiny diamonds in the laboratory by heating and compressing carbon-rich materials — but not coal. Synthetic diamonds such as these are used in dental drill bits and for industrial uses (such as oil-well drilling) where the superior strength of diamonds is beneficial.

Chapter **7**

Recognizing Rocks: Igneous, Sedimentary, and Metamorphic Types

I f you are new to looking at rocks, you may think they are all basically the same — a rock is a rock is a rock. And if you have always been fascinated by rocks and pebbles, you may think there is an endless array of different types. The truth lies somewhere in between.

Many different sizes, colors, shapes, and textures of rocks exist. But all rocks can be sorted into three major categories based on how they are created: igneous, sedimentary, and metamorphic. Within each category, rocks are further sorted by their *composition* (what they are made of) or *texture* (what they look like).

In this chapter, I describe not only the different rock types and how they are categorized but also the processes that create them, such as volcanism, weathering, and metamorphism. You'll never look at a rock the same way again!

Mama Magma: Birthing Igneous Rocks

In order to create an igneous rock, you first need magma, or melted rock. Scientists have identified three ways to melt rock: release pressure, add volatiles, or add heat. In this section, I explain the processes by which rock is melted, describe the specific locations on Earth where these processes occur, and explain how magma can change, or evolve, before cooling into an igneous rock.

Remembering how magma is made

Magmas are created when the conditions of heat and pressure are just right for melting rocks into liquid. There are three specific ways to create conditions just right for melting rock, which I describe in this section.

>> **Decompressing hot rock:** Mantle rock just below the oceanic crust is hot enough to be liquid, but remains solid because of the pressure applied by the crustal rock above it. In some places, such as mid-ocean ridges (see Chapter 9), that pressure is relieved, or removed, as the plates move apart. The removal of pressure, or *decompression*, allows the heated rock to expand and transform into liquid rock, or magma.

>> **Lowering the melting point:** Another way to melt rock is to add volatiles, or compounds such as water and carbon dioxide. These are called volatiles because at most temperatures they prefer to be in a gaseous state. When volatiles are added to a rock that is heated, but not hot enough to melt, they react with the minerals and cause melting — effectively lowering the melting temperature of the minerals in the rock. This melting process is called *flux melting*.

 Magmas created by flux melting occur at subduction zones. (I explain subduction in Chapter 10.) It is in these locations that the subducting oceanic plate carries water and carbon dioxide deep below the crust, where they are released into the overlying asthenosphere and crustal plate, causing the rocks to melt.

>> **Transferring heat:** The simplest way to melt rock is to add enough heat that, without volatiles and without decompression, melting still occurs. This is called *heat transfer* melting and is observed at volcanic hotspots such as Hawaii. In such locations a superheated stream of mantle rock, called a *mantle plume*, brings heat from deep within the earth up to the bottom of the crustal rocks and they melt into magma.

Classifying melt composition

Where and how a melt forms determines the composition of the magma. Geologists classify magma (and igneous rocks) by their relative composition of silica-rich minerals. There are four categories:

>> **Felsic:** Contains 65 to 75 percent silica by weight

>> **Intermediate:** Contains 52 to 65 percent silica by weight

>> **Mafic:** Contains 45 to 52 percent silica by weight

>> **Ultramafic:** Contains less than 45 percent silica by weight

A magma created at a mid-ocean ridge, through decompression of the mantle rock below, has low amounts of silica-rich (felsic) minerals and high amounts of iron-rich (mafic) minerals because Earth's mantle is composed of relatively low silica. Magma with this composition is called a *mafic* or *basalt magma.* (Keep reading to find out which igneous rocks form from this type of melt.)

The magma created at subduction zones due to flux melting has high amounts of silica because the crustal rock that melts is composed of a large percentage of silica-rich minerals. This type of magma is called a *rhyolite* or *felsic magma.*

Heat-transfer melting creates a variety of magma compositions, depending on the type of crust being heated and melted. For example, in Hawaii, the heat-transfer melting produces mafic, or iron-rich, magma; at continental hotspot locations such as Yellowstone National Park in Wyoming, the magma formed by heat-transfer melting is more silica-rich — because the rocks that are melting have more silica-rich minerals.

The initial melting of *source rock* (the originally melted rock) is only the beginning of a melt's journey to make a rock. As the magma moves through the crust it adds new materials through partial melting (keep reading for details on this in the next section, when I explain Bowen's reaction series) as well as through mixing with other magmas and assimilating, or adding, materials from inside the volcanic structure before it erupts. I explain how the melting and cooling of different minerals in a magma can change its composition in the next section.

Reacting in sequence: Bowen's reaction series

The minerals that form rocks by cooling from magma (or lava) change from liquid to solid at different temperatures. A scientist named Norman L. Bowen observed that the igneous minerals that form as a melt cools do so in a particular order, some *crystallizing*, or becoming solid, at much higher temperatures than others. He observed this phenomenon in laboratory experiments and related it to his observations of rocks. He created a reference chart for the sequence of *mineralization* (formation of minerals) in igneous rocks, which is now called *Bowen's reaction series*, shown in Figure 7-1.

Bowen's Reaction Series

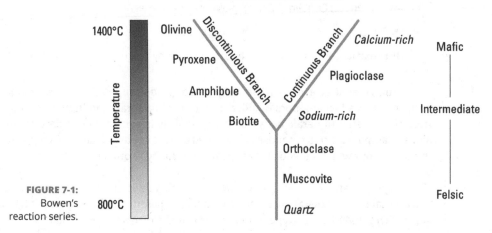

FIGURE 7-1:
Bowen's
reaction series.

TIP

Although the crystallization temperature of minerals was originally studied to understand the formation of igneous rocks, it also helps geologists understand the process of *partial melting* — where only some of the minerals in a rock melt and are added to a magma. It may help to think of it as a phase change temperature since the crystallization temperature is the same as the melting temperature for any mineral. When the heat is increasing the mineral melts at that temperature, but when the heat is decreasing, the mineral crystallizes at that temperature.

TIP

To visualize mineralization through Bowen's reaction series, imagine a soup pot full of chocolate chips, marshmallows, peanut butter, and walnuts. As this soup pot heats up, the "particles" begin to melt. Each item has a different melting point: The marshmallow melts first, then the peanut butter, then the chocolate chips, and finally the walnuts. (It has to be really hot to turn walnuts into liquid!) When the heat is really high, all these items melt into a liquid with no chunks — just a smooth, flowing liquid mix of chocolate, marshmallow, peanut butter, and walnuts. This liquid is like a magma: a fluid mix of melted rock materials.

When you turn the stove off, the pot begins to cool down. Imagine that as it cools, the stuff in the soup pot reforms its original shape. These items will be created in the reverse order in which they melted — in other words, the stuff that melted first will re-form last.

The first item to re-form is the walnuts because they were the last to melt (walnuts have the highest melting point). As the temperature cools back to below that melting point, the walnuts reshape. After the walnuts are re-formed, the elements that make up the walnuts are taken out of the soupy mix. The next items to form must be created from what remains. The chocolate chips mineralize out of the cooling liquid next. Now, all that is left in the liquid is the stuff to make peanut butter and marshmallows. After the peanut butter is resolidified, the only elements left in the soup are the ones that create marshmallows.

REMEMBER

Magmas cool in this same way: As the minerals with the highest melting points are cooled and crystallized, they remove elements from the magma so that only certain elements are left, and these elements can create only certain minerals when they bond together. This process is called *fractional crystallization.*

Fractional crystallization explains how a single magma can produce rocks that can be ultramafic, mafic, intermediate, and felsic. For example, silicate minerals (felsic minerals) are the last to crystallize, so the rocks formed at the end of the cooling (such as granite) are chock full of silica-rich minerals. That's because elements such as iron and magnesium have already been removed to form crystals that solidify at higher temperatures, creating more mafic minerals earlier in the cooling process.

TECHNICAL
STUFF

In Bowen's reaction series there are two lines — one called the Discontinuous Series and one called the Continuous Series. In the Discontinuous Series, a completely different mineral forms at each temperature. In the Continuous Series, however, the same mineral, called *plagioclase*, forms at all temperatures, but at higher temperatures the plagioclase has more calcium in its structure, and at lower temperatures it has more sodium. (Visit Chapter 6 for details on how different elements can be used to build the same mineral.)

Evolving magmas

As a magma moves upward through the crust it may experience cooling, reheating, mixing with another melt, and other changes to its composition. Through these processes a magma *evolves,* or changes, over time. This changing of a magma that produces different igneous rocks is called *magmatic differentiation.*

In reality, magmas don't always start from completely melted rock materials and then cool in sequence according to their melting point as Bowen's reaction series describes. Reality is messy! Realistically, as a magma is created, the minerals that melt first are the ones with lower melting points and high silica content (such as quartz and orthoclase), but the magma may not get hot enough for the mafic minerals to melt, so they remain in rock form. This process is called *partial melting*, which commonly changes magma composition, continuously adding low-temperature-melting-point minerals as it moves upward through the crust. By the time a magma cools, or erupts, partial melting has led it to become more felsic than it was when it first began to melt.

REMEMBER

The evolution of magmas occurs deep in the earth's crust. After a magma has erupted (becoming lava), its composition doesn't change, but the size of the crystals may be different depending on how quickly it solidifies into rock. (In the next section, I explain how the size of the crystals relates to igneous rock texture.)

Which minerals are melted into a magma also determines how fluidly the magma flows. This characteristic of magma is called its *viscosity*. Viscosity is how strongly a fluid resists flowing: Consider peanut butter and water. If you put peanut butter on a plate, it will not flow outward toward the edges — it has high viscosity; however, if you pour water onto a plate, it will flow easily to every edge — it has low viscosity. A felsic magma with high amounts of silica is more viscous; it flows more slowly than mafic magma with low amounts of silica. The viscosity of magma plays an important role in building volcanic structures: viscous, felsic melts create explosive stratovolcanoes, whereas mafic melts with low viscosity create flowing shield volcanoes. (For more on volcanic structures, see after the next section!)

TIP

Magmatic differentiation, mixing, and partial melting explain how so many different types of igneous rocks exist and may help you understand that igneous rocks cannot always be neatly categorized as one type or another; sometimes they fall somewhere in between two categories. For more details on the wide variety of igneous rocks, keep reading!

Crystallizing one way or another: Igneous rocks

Igneous rocks form when molten (melted) rock cools; it changes from a liquid to a solid, or *crystallizes.* If the cooling process occurs underground, the melted rock (which is called *magma*) cools to form *intrusive* or *plutonic* igneous rocks. If the process occurs above ground, such as from a volcanic eruption, the melted rock (which is called *lava*) cools to form *extrusive* or *volcanic* rocks.

REMEMBER

If you find the difference between intrusive and extrusive rocks confusing, keep in mind that the names are telling you where the rocks are formed: *Intrusive* rocks are formed below the surface, or *in* the earth, whereas *extrusive* rocks are formed above the surface, where they have *ex*ited the earth.

In the sections that follow, I show you how various igneous rocks are classified by their mineral composition and their texture, or crystal size. I then provide some basic information about volcanoes — the geographic features that spout lava and create extrusive rocks. Finally, I introduce some underground, or intrusive, igneous rock features.

Classifying igneous rocks

Igneous rocks are *classified*, or named, according to their composition (mineral content) and texture (the size of the crystals). The elements in the melt (magma or lava) determine the mineral content (and color) of the rocks that it creates. How quickly the molten material cools into a solid (rock) determines the texture.

Counting silicate minerals in igneous rocks

The composition, or mineral content, of an igneous rock is determined by the elements that exist in the liquid from which it crystallizes. Most magmas are composed primarily of silica (formed by silicon and oxygen) with portions of other elements such as aluminum, calcium, magnesium, potassium, sodium, and iron. The relative amounts of these elements in a magma determine which minerals are created, as well as the color of the igneous rock.

The terms listed here are the same ones used to describe the composition of magmas in the previous section. They are also used to indicate the differences in composition of igneous rocks because the composition of the magma determines the mineral content of the rock it produces:

>> **Felsic:** These rocks contain more than 65 percent silica, with portions of other elements such as aluminum, potassium, and sodium. Felsic rocks are usually light in color, dominated by quartz, potassium feldspar, and sodium-plagioclase feldspar minerals. Examples of felsic rocks are granite and rhyolite.

>> **Intermediate:** This category of rocks contains between 55 and 65 percent silica with portions of other elements such as aluminum, calcium, sodium, iron, and magnesium. Intermediate rocks are a nearly equal mix of light- and dark-colored minerals, such as amphibole and various plagioclase feldspars. Examples of intermediate rocks are andesite and diorite.

>> **Mafic:** Mafic rocks have between 45 and 55 percent silica with portions of other elements such as aluminum, calcium, iron, and magnesium. They are usually fairly dark in color, with minerals such as pyroxene and calcium-plagioclase feldspar. Examples of mafic rocks are basalt and gabbro.

>> **Ultramafic:** With very low amounts of silica (less than 45 percent) mixed with magnesium, iron, aluminum, and calcium, ultramafic rocks are dark and not very common. They contain olivine and pyroxene minerals, and examples include komatiite and peridotite.

Table 7-1 provides a quick overview of the composition, color, and names of various igneous rocks.

TABLE 7-1

Classification of Igneous Rocks

	Felsic	Intermediate	Mafic	Ultramafic
	Felsic	Intermediate	Mafic	Ultramafic
% Silica	>65%	55–65%	45–55%	<45%
Primary Minerals	Quartz, potassium feldspar, sodium-plagioclase feldspar	Amphibole, sodium- and calcium- plagioclase feldspar	Pyroxene, calcium-plagioclase feldspar	Olivine, pyroxene
Color	Light (white, pink)	Equal parts dark and light (black and white/pink)	Mostly dark	Dark, sometimes green
Names of Intrusive Rocks	Granite	Diorite	Gabbro	Peridotite
Names of Extrusive Rocks	Rhyolite	Andesite	Basalt	Komatiite (rare)

TIP

The terms *felsic, intermediate, mafic,* and *ultramafic* can also refer to the minerals that form igneous rocks. For example, minerals made of silica, such as quartz, may be called felsic minerals.

Observing textures of igneous rocks

Texture is the appearance of the rock — and to some extent how it feels when you touch it. It describes the size of the individual minerals in the rock. The texture of an igneous rock provides clues to how and where the rock was formed: specifically, how long it took to cool from the magma (or lava). As a general rule, igneous rocks that form underground have larger minerals than ones that form aboveground, because temperatures cool more slowly underground than at the surface.

Following are the terms commonly used to describe the texture of igneous rocks. For each term, I indicate whether it usually describes an intrusive (I) or extrusive (E) rock:

>> **Phaneritic** (I): An igneous rock with crystals that are large enough to see without a microscope is called *phaneritic*. This texture is created when the rock forms deep underground from a magma that cools very slowly. The slow cooling time allows the crystals to grow fairly large, transforming liquid elements into solid mineral form (crystals) one atom at a time over a long

period. The result is that all the crystals are large enough for you to see, and they are all approximately the same size. The crystals may be different colors or shapes depending on the composition of the magma and the minerals that form during cooling.

>> **Aphanitic** (E): An igneous rock with *aphanitic* texture is created when lava cools very quickly — so quickly that the minerals don't have time to grow and are very tiny. You can see the individual mineral crystals in an aphanitic rock only through a microscope.

>> **Porphyritic** (I, E): A *porphyritic* igneous rock is created by slow cooling followed by quick cooling of a magma. A magma begins to cool slowly, and then its environment changes (perhaps it is pushed up closer to the earth's surface or erupts as lava) so that it suddenly cools much more quickly. The resulting rock will be composed of some large crystals (that grew during slow cooling) mixed in with smaller crystals (that cooled quickly). Large crystals trapped within a mix, or *matrix,* of smaller crystals is a porphyritic texture. In a porphyritic igneous rock, the large crystals are called *phenocrysts.*

>> **Pegmatite** (I): Some intrusive igneous rocks form near (but still below) the surface under conditions of low temperatures, with high amounts of water mixed into the magma. In a pegmatite-forming magma, the water helps the ions (see Chapter 5) move around and form large crystals. The result is a rock composed of *very* large crystals with no matrix of smaller crystals around them; it's just a mash of giant crystals. This type of igneous rock is called a *pegmatite.*

>> **Glassy** (E): An igneous rock has a *glassy* texture if it cools extremely fast from a lava flow — so fast that the atoms don't have time to form crystals at all, resulting in a smooth, glassy surface. This texture is most common in lavas that have a very high amount of silica in them. (Silica is the same molecule that makes the glass in your windows.)

>> **Vesicular** (E): A rock full of holes (like a sponge) is described as *vesicular.* As the rock cools, gas bubbles are trapped in it. Eventually, the gas escapes, leaving holes called *vesicles* in the rock. This characteristic is most common in rocks that form from volcanic eruptions.

>> **Pyroclastic** (E): When a volcano erupts, other materials such as fragments of rocks from the volcano walls and ash may erupt along with the lava. Rocks that form from these erupted materials are called *pyroclastic.* Pyroclastic rocks may look like sedimentary rocks (which I describe in a later section of this chapter), but they are still igneous — composed of volcanic rock materials. If the fragments are small (less than 2 millimeters), the rock is called *tuff.* If the fragments are large (more than 2 millimeters), the rock formed is called *volcanic breccia.*

A BASALT BY ANY OTHER NAME: PAHOEHOE AND A'A

The content of a lava — which elements are in the liquid and whether there are dissolved gases — gives the lava certain characteristics, such as how fluidly it flows. Hawaii's volcanoes called Mauna Loa and Mauna Kea provide a safe location to observe basalt magma flows up close. From observations of Mauna Loa and Mauna Kea, *volcanologists* (scientists who study volcanoes) distinguish two very different types of basalt magma flow.

Pahoehoe (pah-hoi-hoi) is a wrinkled, ropy-looking lava (and eventually basalt rock) with a smooth surface; it's shown in the picture below. As the lava with low amounts of silica and very little dissolved gasses flows, the outer surface, exposed to the air, cools and solidifies. However, the inner lava is still flowing, which creates the wrinkles and twists in the cooled crust of the flow.

Another type of lava produced from Hawaiian volcanoes is a'a (ah-ah) lava. *A'a* lava is also composed of basalt but is very different from pahoehoe. Instead of fluidly flowing, a'a is blocky, rough, and jagged. As it flows forward it looks, and sounds, like an advancing pile of basalt rubble. A'a lava usually has a high amount of gas dissolved in it, and as it cools the gas escapes, leaving sharp spiny projections of basalt.

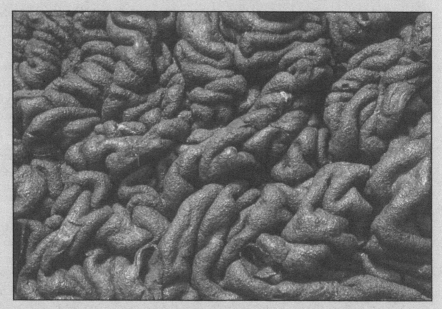

© *Universal Images Group/Getty Images*

These two types of basalt lava were first observed and named in Hawaii, but now scientists use the terms pahoehoe and a'a to describe basalt lava flows with these characteristics no matter where they are located.

Studying volcanic structures

Extrusive igneous rocks are also called *volcanic* rocks. That's because they are created from lava that erupts from volcanoes. *Volcanoes* are the geologic structures that result from magma forcing its way up through the crustal layer of the earth and reaching the surface. When the magma reaches the surface, it erupts as lava.

In this section, I give you an overview of volcanoes so you can understand how extrusive igneous rocks are created.

Spotting volcanic features

REMEMBER

Lava is not the only material that erupts from volcanoes. Many volcanic eruptions also release gases, ash *(tephra)*, and fragmented rock particles *(pyroclastics)* into the air. These erupted materials build up around the opening of the volcano and create volcanoes of different shapes and sizes.

Volcanoes have certain features, which I describe here and illustrate in Figure 7-2:

>> **Magma chamber:** Deep beneath the surface of the earth (about 60 kilometers or 37 miles), magma is created and held in the magma chamber until enough pressure builds to push it toward the surface.

>> **Pipe:** Magma moves from the magma chamber up to the surface through a *pipe.*

>> **Vent:** Magma exits the pipe onto the surface at a *vent.* Vents that erupt only gases (no lava) are called *fumaroles.*

>> **Crater:** The vent usually opens into the *crater,* which is a depression created at the top of a volcano from the collapse of surface materials inward when the magma chamber is emptied after an eruption.

>> **Caldera:** A *caldera* is a very large crater formed when the top of an entire volcanic mountain collapses inward.

>> **Dome:** A volcanic *dome* is created when erupting materials cover the vent, creating a dome-shaped feature that grows as gas and magma continue to fill it, until the pressure within forces another eruption.

>> **Cone:** A volcanic *cone* is a mountain-like structure created over thousands of years as the volcanic lava, gas, ash, and pyroclastics (rock fragments) spill out onto the surface.

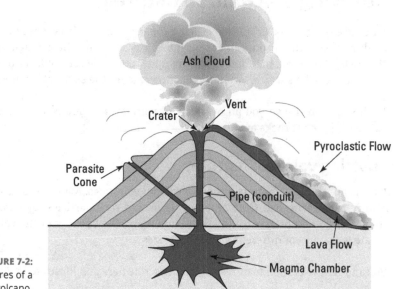

FIGURE 7-2:
Features of a
volcano.

Ash Cloud

Vent

Crater

Pyroclastic Flow

Parasite
Cone

Pipe (conduit)

Lava Flow

Magma Chamber

Distinguishing three types of volcanoes

You may be familiar with the image of a volcano as a tall mountain with smoke exiting from the very top. However, volcanoes can take different shapes depending on the materials that create them. Here I describe three of the most common types of volcanoes.

SHIELD VOLCANOES

Shield volcanoes usually form from basalt lava that erupts through a vent in the ocean floor (though they can form on continents as well). Shield volcanoes are the largest volcano type created by erupting magma. For example, Mauna Loa, Hawaii, spans more than 50 kilometers (31 miles) across.

Figure 7-3 illustrates the major features of a shield volcano. Over time the erupting basalt builds up, creating islands that are low and spread out over a wide area forming a shield shape. As the shield grows, lava flows from the original vent, as well as other vents, creating *flank eruptions* (or *fissures*) along the slopes of the shield. Because the magma of a shield volcano is created from mantle rocks, the melt is mafic and has low viscosity, which allows it to flow outward and create the shield shape. The resulting igneous rocks (such as basalt) are also primarily composed of iron-rich or mafic minerals.

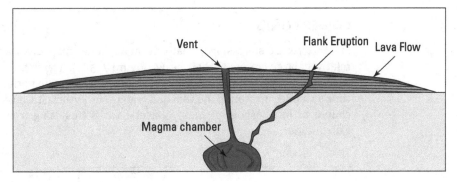

FIGURE 7-3:
Features of a
shield volcano.

STRATOVOLCANOES

The type of volcano you probably imagine when you think about a volcano is a *stratovolcano*, *dome volcano*, or *composite cone*. Most of these type of volcanoes are found around the Pacific Ocean, including the recently eruptive Mount St. Helens in Washington State. Their magma is created as continental crust subducts and causes flux melting in the continental crust. (See Chapter 10 for details on how subduction occurs.) As a result, their magmas have a higher silica content and therefore higher viscosity than shield volcanoes. The outcome is that the shape of the stratovolcano is more upright, with a tall peak in the middle.

Stratovolcanoes have high peaks, steep slopes, and small craters as illustrated in Figure 7-4. They span up to 10 kilometers (6 miles) across — much smaller than shield volcanoes. The peak is created by layers of *andesite* (an igneous rock with high silica content) and pyroclastic materials built up from previous eruptions.

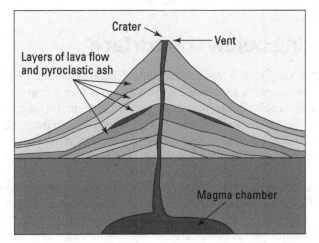

FIGURE 7-4:
Features of a
stratovolcano or
composite cone.

CINDER CONES

Cinder cones are steep sided, relatively small, cone-shaped volcanic features with relatively large craters at their peaks. Figure 7-5 illustrates the common features of a cinder cone volcano. Eruption of mafic pyroclastic material creates a cinder cone as it piles up around the vent and crater. The pyroclastic material may include chunks of lava, ash, and glassy vesicular rocks (resulting from gas-filled lava) called *scoria*.

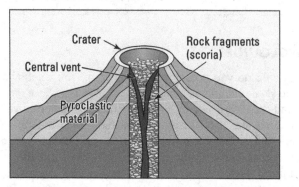

FIGURE 7-5: Features of a cinder cone volcano.

Cinder cones are the smallest eruptive volcanic features — usually less than 1 kilometer (.6 mile) across. After one erupts, its pipe may close and never erupt again. Commonly, you see cinder cones on the side of larger volcanoes, as *parasite cones* of a bigger volcano, such as a shield volcano or stratovolcano (refer back to Figure 7-2). They also commonly occur in groups, or *fields*, such as on the flanks of Mauna Kea (a shield volcano) in Hawaii.

Looking below the surface

Volcanoes and volcanic structures are the surface features of erupted materials. But magma doesn't always reach the surface before it transforms into a solid. When magma solidifies below the surface of the earth, it creates intrusive igneous features. The processes that create these features often occur alongside the processes of volcanism I describe in the previous section. A single magma may produce the multiple volcanic and plutonic features found in a region.

Plutons are solidified magma underground that form bodies of igneous rock below the surface. The shape of a pluton is described as a somewhat rounded or full shape. Tabular igneous features form when magma fills cracks in preexisting rock and the resulting rock bodies are flat or linear in shape and are called *sills* (when they are horizontal) and *dikes* (when they are vertical).

If the igneous feature cuts through layers of rock (usually sedimentary rocks, which I describe in the next section), it's called *discordant*. If it fills cracks parallel to preexisting sedimentary layers, it's called *concordant*.

Figure 7-6 illustrates the most common intrusive igneous features, which I describe here:

>> **Dike:** *Dikes* are tabular, discordant features created by magma filling narrow cracks or fractures in crustal rocks.

>> **Sill:** *Sills* are tabular concordant features created when magma fills space horizontally (usually).

>> **Laccolith:** *Laccoliths* are created by a horizontal flow of magma between sedimentary rocks that are close to the surface and cause an upward bulge or dome shape to the earth's surface.

>> **Pluton:** *Plutons* are rounded or blob shaped intrusions that form when a magma chamber cools and solidifies. Plutons can be from 10 meters to 10s of kilometers wide.

>> **Batholith:** A below-ground igneous feature that is more than 100 square kilometers (62 square miles) in size at the surface of the earth is called a *batholith.* These massive, discordant intrusive igneous features commonly make up the core of large mountain systems, such as the Sierra Nevada mountains in California.

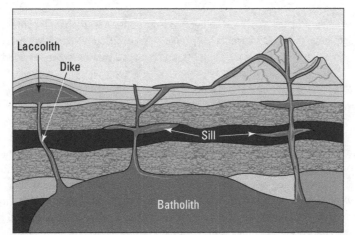

FIGURE 7-6:
Types of intrusive igneous features.

Merging Many Single Grains of Sand: Sedimentary Rocks

Another type of rock, *sedimentary rock*, is much more common on the surface of Earth than igneous rock. Sedimentary rocks are composed of pieces of other rocks. When any rock is exposed to the elements — sunlight, wind, and water — it is eventually broken down into smaller particles. These particles are moved around the surface of the earth by gravity, water, ice, and wind (see Part 4 for details). They eventually settle somewhere, are glued together, and become a sedimentary rock.

Okay, maybe the whole thing isn't quite that simple. In this section, I describe the processes that create sediment particles from existing rock, how the transportation of sediment particles changes their shape and size, and how those characteristics are used to *classify* (categorize) different sedimentary rocks.

DIGGING IN THE DIRT: SOILS

All this talk about sediments may have you thinking about dirt. Dirt as you know it is somewhat different than sediments. A more technical name for dirt is *soil.* Soils develop from the interaction of sediment with air, water, organic materials (plants and animals), and bedrock. Soils are the sediments that support plant growth by providing mineral nutrients, as well as bringing water and air to the plant roots.

As illustrated here, soils are categorized into zones based on their content. This sequence of zones is called a *soil profile.* A soil profile develops as air and organic materials move down into the sediments, and minerals weathered and eroded from the underlying bedrock move up through the sediments. Water also moves from the surface into the sediments, and it flows through the sediments as ground water.

The very top of a soil profile is the O horizon, composed of *humus:* a layer of fluffy, air-filled sediments mixed with high amounts of carbon resulting from decaying plant materials. Below that is the A horizon, which still has high levels of organic material but contains more mineral sediments than the O horizon. Below the A horizon is the B horizon, a layer of sediments with very little organic material and high amounts of weathered mineral sediments. At the bottom of the soil profile is the C horizon, made of weathered bedrock sediments with no organic materials. Below the C horizon is solid bedrock.

Soil classification, or the naming of different types of soil, is extremely complex. The scientific names of soil types try to describe the bedrock that is being weathered, as well as the climate (tropical and wet, or dry like a desert) and the stage of development the soil is currently in.

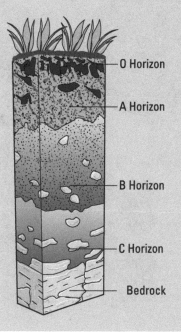

- O Horizon
- A Horizon
- B Horizon
- C Horizon
- Bedrock

Weathering rocks into sediments

Rocks at the earth's surface are *weathered*, or changed by contact with water, heat, wind, ice, and other natural processes. Everything exposed to the "weather" (water, wind, and so on) is weathered; you may notice that anything you leave outside for a long time becomes faded, scratched, chipped, or rusted. These changes are the result of weathering, just like rocks experience.

REMEMBER

When rocks are weathered, they change shape and sometimes composition. Weathering changes the parent rock in one of two ways: by breaking it into smaller particles called *sediments* or by changing its mineral composition through ion exchange (see Chapter 5). I explain these different weathering processes here.

Chipping away: Mechanical weathering

Mechanical weathering (also called *physical weathering*) changes the shape and size of a rock, breaking it down into smaller pieces without changing its chemical composition. Through mechanical weathering, rocks are broken down into sediment particles. This process is illustrated in Figure 7-7.

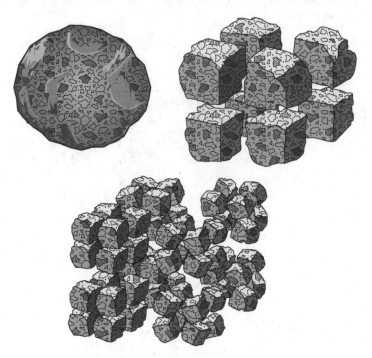

FIGURE 7-7:
Mechanical
weathering of a
rock into
sediment.

There are four common types of mechanical weathering:

>> **Frost wedging:** This type of weathering is most common in cold mountainous regions and occurs when water seeps into cracks in a rock and freezes — expanding and pushing the rock apart, or breaking off a piece of it. Usually, frost wedging takes a few cycles of freezing and thawing so that the ice first expands the cracks in the rock before eventually breaking it apart.

>> **Thermal expansion:** Layers of a rock may also break off due to expansion from heating. This type of mechanical weathering is called *thermal expansion*. In the case of thermal expansion, usually only the outer layer of the rock breaks off. Rocks do not absorb heat very well, so the minerals in the outer layer heat up, expand, and break off, while the inner portions of the rock stay cool.

>> **Unloading:** Exposed igneous rocks break apart in a process called *unloading, sheeting,* or *exfoliation.* As plutons (explained earlier in the chapter) are exposed, the outer layers of rock are released from the great pressure of being buried deep in the earth. In response to this pressure release, the minerals expand and separate from the underlying rock in a *concentric* or *domed* pattern that is illustrated in Figure 7-8.

» **Abrasion:** The most common mechanical weathering is *abrasion,* or the scraping of rocks against each other. In Part 4, I explain the different ways abrasion occurs when I explain geologic surface processes.

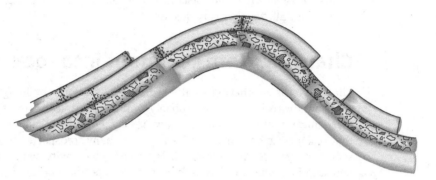

FIGURE 7-8:
Weathering by exfoliation.

Reacting with water and air: Chemical weathering

Chemical weathering changes the mineral composition of the rock, removing or exchanging ions from the minerals, or even dissolving them completely. (I explain ion exchange in Chapter 5.) Chemical weathering affects only the surface of a rock. As the rock is *eroded* or broken down by mechanical weathering, it becomes more vulnerable to chemical weathering because more surface area is exposed.

Three types of chemical weathering can occur:

» **Dissolution:** *Dissolution* occurs when water removes ions from a mineral. For example, the mineral halite (salt) is easily dissolved in water, so rocks with halite minerals are easily weathered by dissolution.

» **Oxidation:** Oxygen in the atmosphere (the same oxygen you breathe) can bond with some ions in a process called *oxidation.* The rust you see on metal surfaces is the result of the iron in the metal bonding with the oxygen in the air, or *oxidizing.* This same process can occur with iron-rich minerals that are exposed to oxygen in the air.

» **Hydrolysis:** *Hydrolysis* is the exchange of a hydrogen or hydroxide ion (the H and OH found in water, or H_2O) for an ion in a mineral. This exchange most often occurs when silicate minerals such as feldspar (see Chapter 6) are exposed to water and transform into clay minerals.

Water is the most important agent of chemical weathering. Rates of chemical weathering are determined by the temperature and amount of water available. The result is that climate, and therefore geography, play an important role in how quickly rock is eroded by chemical weathering. Rocks in cold, dry, polar regions weather very slowly while rocks near the equator, with warm temperatures and regular rainfall, weather very quickly.

Changing from sediment into rock

The results of weathering — the sediments — are the building blocks of most sedimentary rocks. You are probably familiar with sediments that result from mechanical weathering such as sand or clay. Both these sediments and the ions that result from chemical weathering are usually transported before becoming a sedimentary rock. The sediment is transported by gravity, wind, water, or ice to a new location. I explain how each of these agents transports rock and sediment in Part 4.

If sediment is transported by gravity or ice, particles of all different sizes will be moved at the same time, resulting in an *unsorted* or *poorly sorted* mix of sediments. This means large boulders, sand grains, and tiny clay particles may all be together.

However, if the sediments are transported by water or wind, they are *sorted* by size. Wind and water, depending on speed, can carry only particles of certain sizes (as I discuss in Chapters 12 and 14). The result is that a *well-sorted* mix of sediment has particles that are all the same size. Figure 7-9 shows the difference between poorly sorted and well-sort sediments.

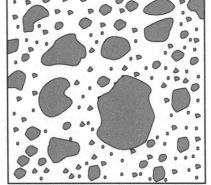

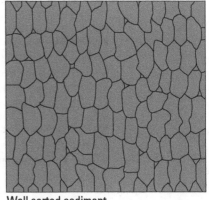

FIGURE 7-9:
Poorly sorted and well-sorted sediments.

Poorly sorted sediment

Well sorted sediment

REMEMBER

The distance that sediment travels and its mode of transportation leave visible changes in the particles, sometimes rounding them or scratching them. These changes, along with the size of the particles, give a sedimentary rock its texture and grain size. These are the characteristics geologists use to classify and name sedimentary rocks.

After the sediments are transported and *deposited* (left in place) for a long period of time, they go through various processes of *diagenesis,* or change.

The first stage of diagenesis is burial of the sediments. When the sediment is buried, things start to heat up, and the pressure is increased. These conditions lead to *lithification:* the process that turns sediments into rock. It starts with *compaction,* where the pressure and weight of overlying sediments presses down, squeezing any extra space out of the layers of sediment.

Following compaction is *cementation,* or gluing together the sediments into solid rock. The individual grains are cemented together when water moves through the sediments, carrying dissolved elements. These elements from minerals that fill the tiny spaces between particles and build crystals linking the grains together. Most sedimentary rocks are cemented by calcite, silica, or iron oxide that formed due to chemical weathering. (See Chapter 6 for details on minerals.)

REMEMBER

Sedimentary rock-forming diagenesis occurs at relatively low temperatures (about 150 to 200 degrees Celsius or less) close to the earth's surface. When temps are hotter than that, metamorphic rocks are created. I describe metamorphic rocks in the upcoming section "Stuck between a Rock and a Hard Place: Metamorphic Rocks."

Sizing up the grains: Classifying sedimentary rocks

Sedimentary rocks are classified into two major groups: *detrital* and *chemical.* I describe each group in this section. Generally speaking detrital rocks are the result of physically weathered sediments, whereas chemical rocks are the results of chemically weathered sediments.

Detrital sedimentary rocks

Detrital sedimentary rocks are made of sediment particles and are described as *clastic,* meaning they are composed of broken bits of other rock, or *clasts.* The primary way of classifying and naming detrital sedimentary rocks is by the size of their sediment grains; see Table 7-2.

TABLE 7-2 **Detrital Sedimentary Rocks**

Grain Name and Size	Rock Name
Clay (< 1/256 mm)	Shale
Silt (1/256 mm – 1/16 mm)	Siltstone
Sand (1/16mm – 2 mm)	Sandstone
Gravel to boulders (> 2 mm)	Conglomerate or breccia

REMEMBER

To classify detrital sedimentary rocks, scientists also consider the shape and mineral composition of the grains. The shape of a sediment grain is determined by how it is transported and the distance it travels. Transportation by water or wind leaves grains *rounded* with no sharp edges. Grains that are not rounded are *angular* and have most likely been transported by ice or gravity. As I explain in the previous section, how well-sorted or poorly sorted the grains are also comes into consideration when classifying these rocks.

Both *shale* and *siltstone* are composed of very tiny sediment grains. In the case of *black shale,* the grains are composed of organic matter, such as decayed organisms, as well as minerals. Both shale and siltstone are considered mudstones, though *mud* is a less scientific term for the tiny grain size. You can easily distinguish shale from siltstone because shale splits into very thin layers (called *laminations*) and you cannot see the individual sediment grains without a microscope.

REMEMBER

For particles this tiny to settle out of water and be compacted, the water must not be moving. Therefore, shale and siltstone are often the result of sediments deposited in still water, such as a lake bed or deep ocean.

Sandstones are categorized according to their mineral content. Three major types exist:

» **Quartz sandstones:** These are the most common sandstones and are composed of well-sorted quartz grains. They are commonly white or tan colored.

» **Arkose sandstones:** These contain high amounts of feldspar (eroded from granite). They are often poorly sorted with angular pink or reddish grains.

» **Greywacke sandstones:** These are composed of sediments eroded from volcanic rocks (such as basalts). They have some quartz and feldspar but are poorly sorted, angular, and generally dark in color.

The largest sediment grains create conglomerates and breccias (see Table 7-2). Almost always, these rocks are poorly sorted, with the larger particles held together by a cemented mix of smaller particles. If the sediments are rounded, the rock is a *conglomerate*. If the sediments are angular, the rock is a *breccia*.

No grains at all: Chemical sedimentary rocks

Chemical sedimentary rocks are made not of rock particles but of chemical particles that are the result of chemical weathering. They are created when minerals are dissolved from existing rock into water and carried to the ocean (or a lake). In a body of water, the minerals re-form into solids, or precipitate, creating particles that settle to the bottom and *lithify*, becoming a sedimentary rock. Chemical sedimentary rock particles may be organic or inorganic:

>> **Organic:** Organic sediment grains come from the shells created by an organism during its lifetime using elements dissolved in the seawater. When the organism dies, the shell remains and becomes sediment particles on the ocean floor — subjected to the same changes in size, shape, and sorting that I describe for sediment grains in the previous section. These types of sedimentary rocks may also be called *biogenic* sedimentary rocks.

>> **Inorganic:** Inorganic sediment grains are any minerals that form a chemical sedimentary rock that are not related to a living organism.

Chemical sedimentary rocks are named according to their mineral composition, as I show in Table 7-3.

TABLE 7-3

Chemical Sedimentary Rocks

Composition	Rock Name
Calcite (shells)	Biogenic limestone (coquina)
Silica (diatom shells)	Biogenic chert
Carbon (plants)	Coal
Calcite ($CaCO_3$)	Inorganic limestone (travertine)
Dolomite ($CaMg(CO_3)_2$)	Dolostone
Silica (SiO_2)	Inorganic chert
Halite (NaCl)	Salt evaporite (rock salt)
Gypsum ($CaSO_4 * 2H_2O$)	Gypsum evaporite

Here's a bit of information about each type:

>> **Biogenic limestone (coquina):** These rocks are shells glued together by the mineral calcite as it *precipitates,* or solidifies, out of warm sea water. The shells may be microscopic, visible, or broken into bits. The cementing mineral calcite is easily dissolved by water, so limestone landscapes often form caves or karst (see Chapter 12) such as Mexico's Yucatan Peninsula.

>> **Biogenic and inorganic chert:** This type of rock is created by the crystallization of silica. The crystals are so tiny they can be seen only with a microscope. Most cherts are inorganic, but some are created by microscopic silica-shelled organisms called *diatoms*. When the organisms die, their shells settle to the seafloor and eventually create *biogenic chert.*

>> **Coal:** Coal forms when massive amounts of plant materials and other living matter are compressed and lithified. Most of the coal that exists today is the remains of swamps from many millions of years ago. As the swampy vegetation dies, *peat* forms and is compacted into soft *lignite* and eventually hard *bituminous* coal. The carbon from the organisms is released into the atmosphere when you burn coal for fuel.

>> **Inorganic limestone (travertine):** These rocks occur when the mineral calcium carbonate precipitates out of water. One way this occurs is when groundwater (see Chapter 12) carries dissolved calcium carbonate into caves. As it precipitates out of solution, it creates an inorganic limestone called *travertine.*

>> **Dolostone:** This rock is similar in appearance to inorganic limestone. But instead of the mineral calcite (formed by calcium carbonate), dolostone is composed of *dolomite,* a mineral made of magnesium and calcium carbonate. Geologists are still trying to figure out why massive layers of dolostone were created in the past but do not seem to occur today.

>> **Salt and gypsum evaporites:** These rocks form when water full of dissolved minerals evaporates. As the water disappears, the minerals form crystals. Usually evaporites are made of salt (the mineral halite) or gypsum because these minerals precipitate out of water first. Minerals that are very soluble stay dissolved in water unless all the water disappears, so they more rarely make evaporite rocks.

Searching for sedimentary basins

In order for sedimentary rocks to form, large amounts of sediment must collect somewhere and undergo the process of lithification. The places where sediments commonly collect on the earth's surface are called basins. There are many different types of basins, from mid-continental basins to deep sea basins and everywhere in between! A basin is anywhere that the surface of the earth is low enough to collect sediments, without an outlet to a lower elevation location. Here are a few of the basins identified within the context of plate tectonics:

>> **Continental rift basins:** These are formed at locations where two continental crust plates are stretching apart, sediments will collect within the low area created between the rifting plates.

>> **Passive margin basins:** When the edge of a continent is not a subduction zone, sediments will collect along the edge of the continent in the ocean, where they are deposited by rivers draining off the land.

>> **Intracontinental basins:** In some places the interior of a continent sinks, or subsides enough to create a basin where sediments will collect.

>> **Foreland basin:** After mountain building occurs (due to tectonic plate collision), the mass of material in the newly built mountain will press down on the crust and create a basin on the continental side of the mountains that is deep enough to collect sediment.

Telling stories of the past: Sedimentary structures

When sedimentary rocks form, they capture a picture of the *environment of deposition*. Features of the rocks, called *sedimentary structures*, are clues to the past. Here I describe the most common sedimentary structures:

>> **Beds:** Most sedimentary rocks are deposited in layers, or *beds*. Each bed is created as grains settle through water and deposit together.

>> **Bedding planes:** Layers of similar grain type (size and shape) will be separated by a *bedding plane*, a flat surface where the grains change from one type to another. Sedimentary rocks often break along bedding planes, where the transition of one grain type to another creates a slight weakness in the cement. Sometimes bedding planes can be recognized by a change in color — appearing as stripes in the rock.

>> **Cross-beds:** While most beds are deposited across a flat or horizontal surface, some sedimentary rocks have beds that are at an angle. These *cross-beds* (see Figure 7-10) result from flowing water or wind depositing the sediments in little piles (or dunes, which I describe in Chapter 14).

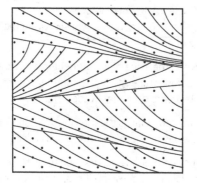

FIGURE 7-10:
Cross-bedding.

>> **Graded beds:** Graded beds (see Figure 7-11) occur when sediments of different sizes settle out of water at different times: the larger grains first, followed by increasingly smaller grains. This gradation indicates that the sediments were carried in a flow of high energy (high enough to pick up larger grains), and then suddenly the water slowed or stopped so that the particles fell to the bottom — the largest and heaviest first.

>> **Ripple marks:** Ripple marks (see Figure 7-12) occur in rocks created at shorelines (like the beach) or in streambeds. You have probably seen the wavy lines of sand created as the water moves back and forth, in and out, at the shore of a lake or ocean. Ripple marks created by this back and forth motion are called *oscillation ripples.* Ripples created by water moving in one direction (such as a creek bed) are called *current ripples.* Ripples can also be created by wind moving sand.

>> **Mudcracks:** *Mudcracks* are created when a standing body of water, such as a lake, dries up. The surface of the mud cracks and splits as it dries (see Figure 7-13). These cracks can be preserved as structures in a sedimentary rock.

Sedimentary structures can be useful in determining if a sedimentary rock layer is still in the position (right-side-up) in which it was originally deposited. Understanding processes that create sedimentary structures helps geologists interpret the history of a rock's formation, as well as the environment at the time it was formed.

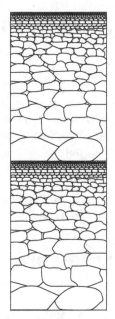

FIGURE 7-11:
Graded bedding.

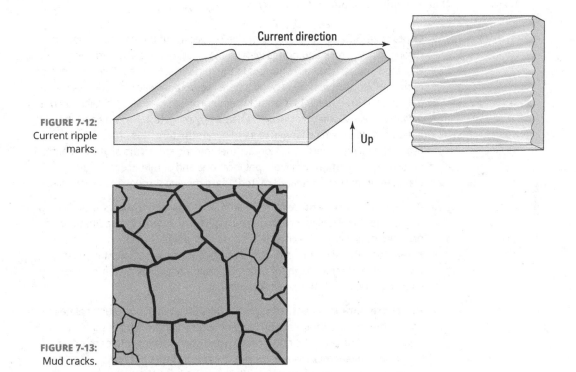

FIGURE 7-12:
Current ripple marks.

Current direction

Up

FIGURE 7-13:
Mud cracks.

Stuck between a Rock and a Hard Place: Metamorphic Rocks

Metamorphic rocks begin as either igneous, sedimentary, or preexisting metamorphic rocks and undergo a major change, or *metamorphosis*. The change is caused by high levels of heat and pressure — levels found deep in the earth's crust, below where sedimentary rocks are formed but not so deep and hot that the rocks are melted into a magma.

In this section, I describe the various stages of metamorphic change, characteristic textures of metamorphic rocks, and how they are classified.

Turning up the heat and pressure: Metamorphism

REMEMBER

Metamorphism changes an original or *parent* rock into a new type of rock. This change happens in one of three ways:

>> **Contact with heat:** *Contact metamorphism* occurs when magma moves up through crustal rock and brings with it high levels of heat. The surrounding rock is heated enough to cause changes in the mineral structures. The transformed minerals create an *aureole* around where the magma was located. This process is illustrated in Figure 7-14. Contact metamorphism may also be called *thermal metamorphism*. Sometimes water is heated by magma in this way and then enters a rock, causing changes to its mineral structure; this process is called *hydrothermal metamorphism*. Contact metamorphism occurs anytime magma is present. It is common near hot spot locations and at mid-ocean ridges or continental rifts where heated materials are moving through the crust.

>> **Burial under rocks and sediment:** *Burial metamorphism* affects rocks buried at great depth (usually more than 6 miles). The rocks are exposed to heat and pressure deep within the crust that cause changes to the minerals in the parent rock. Burial metamorphism is most common in sedimentary basins, where the increasing weight of overlying sediment causes metamorphic change to the deeply buried rocks below.

>> **Direct pressure and heat from plate collisions:** When two crustal plates collide, the result is mountain building or subduction (which I describe in Chapter 10). The pressure and heat of two plates crashing into each other causes *dynamothermal metamorphism*. This type of metamorphism is evident in mountains built by continental collision, as well as along subductions zones — though the subducting plate carries some of the evidence into the mantle!

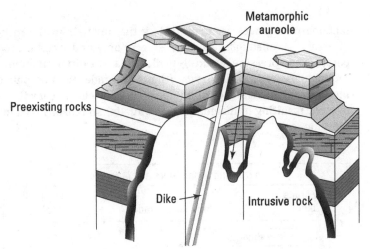

Metamorphic aureole

Preexisting rocks

Dike → | Intrusive rock

FIGURE 7-14:
Contact
metamorphism.

REMEMBER

Both burial and dynamothermal metamorphism affect large areas of crustal rocks and are considered *regional metamorphism*. Contact metamorphism, on the other hand, is very local and affects only the rocks immediately surrounding the heated materials.

Grading metamorphism with index minerals

As parent rocks are exposed to heat and pressure, they begin to change. The degree of change depends on the levels of heat and pressure they experience. The resulting metamorphic rocks are described by their degree of change, or *metamorphic grade*:

» **Low-grade metamorphic rocks retain characteristics of the parent rock.** If they are sedimentary rocks, they may still show signs of bedding planes or other structures. Low-grade metamorphic rocks have been exposed to relatively low temperatures and pressures.

» **High-grade metamorphic rocks look very different from their parent rock.** Rocks exposed to very high levels of heat and pressure change dramatically; their internal structure no longer resembles the original rock and completely different minerals may be present.

In regional metamorphism, large areas of crustal rocks are being subducted or buried and changed. The rocks deeper in the crust are subjected to higher temperatures and pressures than the rocks closer to the surface. The result is that across a region you see rocks of different metamorphic grades corresponding to the increasing degree of metamorphism.

Metamorphic grades are identified by the minerals in the rock because certain minerals — called *index minerals* — form only under certain conditions of temperature and pressure. Figure 7-15 illustrates the different minerals formed as the sedimentary rock shale moves from low-grade metamorphism to high-grade metamorphism. At some point, high enough levels of heat will cause the minerals to melt, resulting in magma and (eventually) an igneous rock instead of a metamorphic rock.

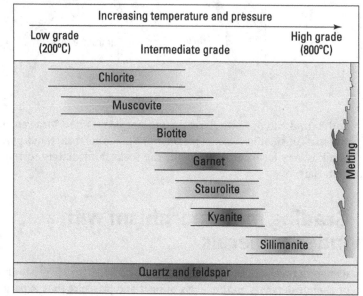

FIGURE 7-15:
Index minerals
in shale
metamorphism.

Between the mineral sheets: Foliation, or maybe not

Pressure is one of the causes of metamorphism. The squeezing of rock minerals under conditions of high pressure forces them to change. Two types of pressure are applied to metamorphic rocks and illustrated in Figure 7-16. *Indirect pressure* pushes on the rocks from all sides, compacting the materials and removing any spaces between crystals or particles. *Direct pressure* comes from two opposite directions and elongates the minerals into parallel layers.

The elongation of minerals by direct pressure creates a texture specific to metamorphic rocks called *foliation*. Foliation occurs when the minerals line up in thin layers under the application of direct pressure. The minerals are compressed or reshaped into long, linear forms, illustrated in Figure 7-17.

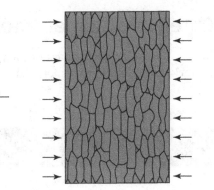

FIGURE 7-16:
Indirect and
direct pressure.

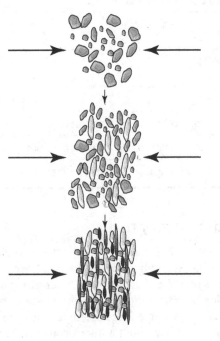

FIGURE 7-17:
The foliation of
minerals by direct
pressure.

However, not all metamorphic rocks are foliated. Rocks that are metamorphosed by contact with heated magma or indirect pressure still experience changes in the organization of the mineral grains, but the minerals do not create sheets or layers. These *metamorphic minerals* are created when the atoms rearrange themselves to form completely different minerals from the parent minerals. Other minerals respond to metamorphism by growing larger (marble is an example).

I describe examples of metamorphic rocks with all these different textures in the next section.

Categorizing metamorphic rocks

Table 7-4 summarizes the classification of metamorphic rocks, including the parent rock, conditions of metamorphism, and texture (or foliation).

TABLE 7-4   ## Metamorphic Rocks

Rock Name	Parent Rock	Metamorphic Conditions (Temperature/Pressure)	Texture
Slate	Shale	Low	Foliated
Phyllite	Shale	Low to intermediate	Foliated
Schist	Phyllite, basalt, greywacke, sand, or limestone	Intermediate to high	Foliated
Gneiss	Schist, igneous rocks, sand, or limestone	High	Foliated
Migmatite	Gneiss	High	Foliated
Marble	Limestone, dolostone	Contact heat or high indirect pressure	Non-foliated
Quartzite	Sandstone, chert	Contact heat or high indirect pressure	Non-foliated
Hornfels	Shale, basalt	Contact heat or low indirect pressure	Non-foliated

Transforming sedimentary rocks

When the sedimentary rock shale, composed of tiny clay particles, is metamorphosed, it first transforms into *slate*. Slate breaks along flat, smooth layers of foliation (which is why slate is used for chalkboards). Under increasing temperature, slate transforms to *phyllite*, which has foliated layers of shiny microscopic mica minerals. When the pressure and temperature are high enough to produce foliated mica minerals large enough to see without a microscope, the rock is called *schist* (pronounced *shist*).

When temperatures reach about 1,200 degrees Fahrenheit, the minerals stop flattening into foliated layers. Instead they try to escape the stress of all that pressure! Certain minerals handle the stress better than others, so the minerals begin to move from high-stress areas to lower-stress areas. The result is *gneiss* (pronounced *nice*): a rock with alternating bands of light (felsic) and dark (mafic) minerals. The separation of light and dark minerals is called *metamorphic differentiation*.

If the pressure and temperatures exceed gneiss-forming conditions, the gneiss begins to melt on its way to becoming magma. When a rock forms from these conditions, it is a *migmatite*. Migmatites are gneisses that have partially melted and then solidified into rock. The minerals are still differentiated into dark and light foliated layers, but they are usually swirled or curvy from all the pressure that nearly melted them into magma. An example of a migmatite is pictured in the color photo section of this book.

Limestones do not proceed through the sequence of metamorphism that I describe for shale. Instead, under conditions of high temperatures and pressure, limestone (and dolostone) minerals are compressed until all the space between crystal grains is squeezed out. The result is a very hard, smooth rock called *marble.* The solid, smooth feature of marble — where the crystals form one continuous body — makes it a great material for sculpting.

Sandstone also creates a very hard metamorphic rock called *quartzite.* Similar to marble, quartzite is formed by compressing all the space from between mineral grains until the crystals are smashed together in one continuous body of mineral grains.

TIP

To illustrate how limestone and sandstone change through metamorphism, imagine holding a handful of frozen blueberries in your hands. While they are frozen, solid, and round, there is space between them. As you squeeze them together — applying pressure and creating heat with your hands — they begin to soften a little, and you can press them closer together. They deform from their round shape to fill all the spaces as the pressure increases. When they are a mass of squished blueberries with no space, refreeze them, and now they resemble marble or quartzite.

Transforming igneous rocks

As basalts are exposed to pressure (but still relatively low temperatures), the minerals transform and become foliated. Low pressures create minerals with a green color, so the metamorphic rock is called greenschist (which has a foliated texture as well as the green color). Exposed to higher levels of pressure, the green-colored minerals transform into blue-colored minerals, creating blueschist. Under increasing temperature and pressure, these schists transform into gneiss, as I describe for sedimentary rocks in the previous section. Intrusive igneous rocks such as granite transform into gneiss rocks as the temperature and pressure force metamorphic differentiation of the minerals into dark and light layers.

Creating hornfels

Hornfels are metamorphic rocks created through contact metamorphism. When magma first moves into a rock close to the surface, it increases temperatures enough to change the mineral composition and texture of the surrounding rocks. However, because no pressure is applied, hornfels are not foliated. Also, they have very small mineral grains because the heating by the magma occurs only for a short time; the minerals don't have time to grow very large before the rock cools again.

TECHNICAL STUFF

Metamorphic rocks form under a wide variety of temperature and pressure conditions deep within the crust. These environments of temperature and pressure are called *metamorphic facies,* and are commonly talked about in reference to the dominant metamorphic minerals produced under those conditions (such as greenschist facies). Of course, two different parent rocks may result in two different rocks forming within a single facies, but a metamorphic geologist will know that they formed under the same conditions of temperature and pressure if they are both described as being of that facies.

Tumbling through the Rock Cycle: How Rocks Change from One Type to Another

If you look back through the previous sections, you may notice that each rock type forms from the remains of a previously existing rock. Each rock is melted and cooled, weathered and deposited, or heated and compressed into something new. There is no beginning or end to this *rock cycle;* it just goes round and round as rocks, sediment, and magma move around the earth's lithosphere.

In Earth's early history (see Chapter 18), molten mantle rock began erupting to the surface, forming the earliest surface rocks. On the surface, these rocks were subjected to weathering. The sediments formed by weathering were deposited in oceans, creating the first sedimentary rocks. These rocks on the ocean floor were then subducted back beneath the surface and exposed to the heat and pressure that creates metamorphic rocks, eventually perhaps being heated enough to melt back into magma. The cycle seems straightforward, but it's not.

Figure 7-18 illustrates the major components of the rock cycle: the rock types and the processes that transform rocks from one type to another.

The rock cycle combines all the processes that create, transform, and destroy the different rock types and shows you how the earth materials that compose these rocks are constantly transitioning from one type to the next.

At each stage of the rock cycle, any rock has multiple paths it may take. For example, a sedimentary rock may be buried, compressed, and transformed into a metamorphic rock; or weathered into sediments; or buried deep in the crust where it is heated and melted into magma and then erupted onto the surface as an igneous rock. As Figure 7-18 illustrates, there are shortcuts in the cycle, and how a rock is transformed depends on where it is on the planet and what forces are acting on it.

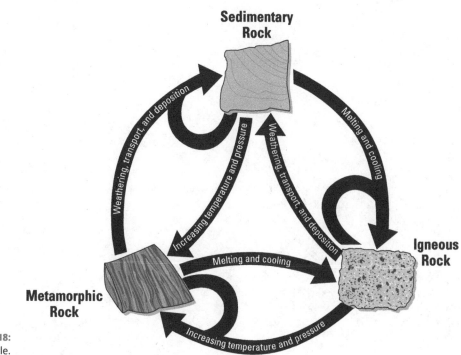

FIGURE 7-18:
The rock cycle.

3

One Theory to Explain It All: Plate Tectonics

Discover the unifying theory of plate tectonics: the single theory that ties all geologic understanding together into one package.

Follow the development of early ideas about continental plate movements.

Learn what happens at the surface and beneath the surface when plates move and how they are driven by multiple forces, including ridge-push, slab-pull, and the welling up of heated rocks beneath Earth's crust.

Chapter **8**

Adding Up the Evidence for Plate Tectonics

Starting in the 1800s, earth scientists of all different types worked to develop theories explaining the different geologic phenomena they observed. Some studied volcanoes, earthquakes, or mountains. Others asked questions about how the continents or oceans were created.

As I explain in Chapter 3, starting in the late seventeenth century, some geologists began to accept that the earth had a long history. (Previously, most people had believed that the earth was only a few thousand years old.) When that shift in thinking occurred, explanations for geologic processes could incorporate long periods of time to create change.

But as geologists developed ideas explaining the different geologic features of the earth, they struggled to explain if, and how, all the geologic phenomena were tied together. What they lacked was a *unifying theory of geology*: a single theory of geologic processes that explained all the phenomena they observed.

In modern geology, such a unifying theory exists: the *theory of plate tectonics*. In this chapter, I describe the development of ideas concerning crustal plate movement, which form the basis of this theory. I also explain how scientists today are gathering evidence to further support the theory.

Drifting Apart: Wegener's Idea of Continental Drift

In the early part of the twentieth century, German scientist Alfred Wegener proposed that all the continents had at one time been connected; they had comprised a single large continent, a *supercontinent,* called *Pangaea.* He proposed that over a long span of geologic time, the location of the continents changed as they separated and moved apart from one another. His idea is known as the *continental drift hypothesis.*

Observations of geography, geology, and the location of fossils across the continents support the idea of an ancient supercontinent. In this section, I describe these various lines of supporting evidence.

Continental puzzle solving

If you look at a map of the world such as Figure 8-1, you will notice that the eastern coast of South America and the western coast of Africa have a very similar shape. Wegener also noticed this fact. He suggested that South America and Africa had once been connected along these coastlines.

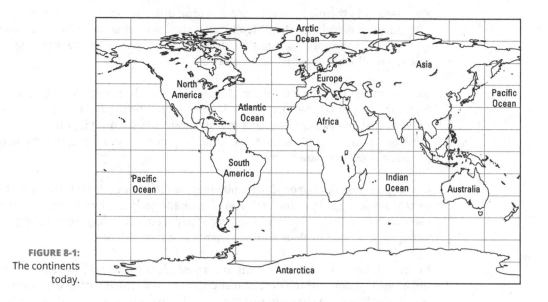

FIGURE 8-1:
The continents today.

Early critics argued against this line of evidence, stating that coastal shorelines are constantly changing and pointing out, correctly, that the fit along the coasts of these two continents was not perfect enough to support Wegener's idea. However, if you include the *continental shelf*, the portion of the continent that continues from the coast underwater for a few miles, when you put the continents together, the fit is nearly perfect, as illustrated in Figure 8-2.

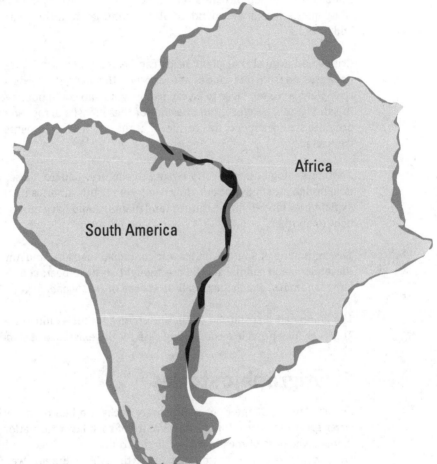

FIGURE 8-2:
South America and Africa connected.

Fossil matching

Another line of evidence supporting the continental drift hypothesis is the similarity of fossils found on multiple continents. For example, fossil remains of a rare, freshwater reptile, the *mesosaur*, are found in both South Africa and South

America. Scientists have also found remains of other reptiles, such as the *Lystrosaurus* and *Cynognathus,* on multiple continents. Because these animals lived on land, it is unlikely that they swam across the oceans that now separate South America, Africa, and Antarctica.

The strongest line of fossil evidence for a supercontinent is from the remains of many different plant types that collectively are known as the *Glossopteris flora* (named for a group of plants that are thought to have lived in ancient swamps). Glossopteris fossils are found on all the continents in the southern hemisphere and in India.

Critics had argued that plant seeds can be spread by wind and water and could have crossed from one continent to another. However, the seeds of this early fern-type plant were too large to be carried by wind and could not have survived travel in salt water. Therefore, the presence of these fossils across so many continents provides strong support for the idea that the southern continents must have been connected.

Another geologist in the early twentieth century, Eduard Suess, proposed that a land bridge must have at one time connected South America to Africa. This would explain how land-living creatures (and plants) could have moved from one continent to another.

REMEMBER

The continents of South America, Africa, India, Australia, and Antarctica linked by these plant and animal fossils are thought to have been connected as a single, large landmass, which Suess call *Gondwana* or *Gondwanaland.*

The combined distribution of these different fossils is illustrated in Figure 8-3. Together they provide a compelling case for the existence of Gondwana.

Stratigraphic stories

As I explain in Chapter 16, sedimentary rocks are laid down in layers and may cover great distances. Due to this feature of rock layer formation, geologists can match sequences of rock layers, or *strata*, from different geographic locations and create a picture of the environment at the time the sediments were deposited. (See Chapter 16 for details on the study of rock layers, or *stratigraphy*.)

The patterns of rock strata on Antarctica, South America, Africa, India, and Australia are illustrated in Figure 8-4. The similarity in rock sequences lends strong support to the idea that these continents were once connected to one another.

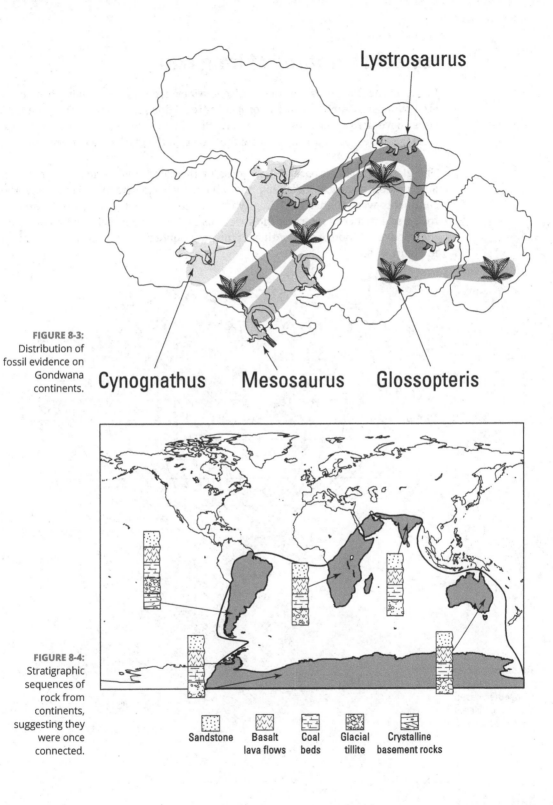

Lystrosaurus

FIGURE 8-3:
Distribution of
fossil evidence on
Gondwana
continents.

Cynognathus   **Mesosaurus**   **Glossopteris**

FIGURE 8-4:
Stratigraphic
sequences of
rock from
continents,
suggesting they
were once
connected.

Sandstone   Basalt
lava flows   Coal
beds   Glacial
tillite   Crystalline
basement rocks

Icy cold climates of long ago

Rocks on the Gondwana continents also show evidence of once having been covered by massive amounts of ice, or *glaciers* (explained in Chapter 13). When glaciers move across rock, they leave scratches in the surface of the rock. These scratched lines, or *striations*, indicate the direction the glacier ice is moving.

The glacial striations observed on rocks of the same time period in Antarctica, South America, Africa, India, and Australia allow geologists to re-create how the continents were connected. Using what they know about glacier ice movement today, scientists can interpret the striations on the ancient rocks and reconstruct the relative position of the continents forming Gondwana. The result is illustrated in Figure 8-5.

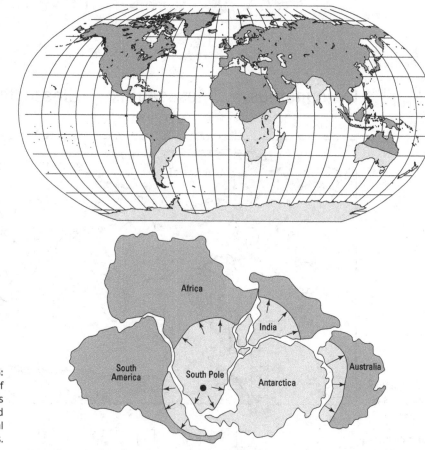

FIGURE 8-5:
Reconstruction of the continents together, based on glacial striations.

Meeting at the equator

All these lines of evidence strongly support the connection of the southern continents into a single, large landmass, but what about the continents of North America, Europe, and Asia?

A South African geologist named Alexander du Toit expanded on Wegener's work by gathering more geologic and fossil evidence during the 1920s and 1930s. He proposed that in the past, Gondwana was situated at the South Pole (based on the evidence for glaciation) and that another large landmass existed near the equator. He called this equatorial supercontinent — composed of North America, Greenland, Europe, and Asia — *Laurasia*. Figure 8-6 illustrates how Laurasia probably appeared.

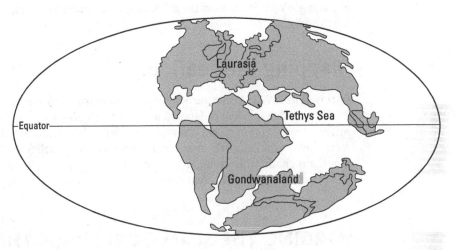

FIGURE 8-6:
The continents forming Laurasia.

You can find more details about the supercontinents Pangaea, Gondwana, and Laurasia in Chapters 20 and 21.

TIP

Searching for a mechanism

Even with all these lines of evidence, scientists were still skeptical of Wegener's idea of continental drift for one very important reason: No one had yet explained *how* the continents moved.

Wegener had proposed that the large continental landmasses moved through the crust of the ocean floor like an icebreaker through ice, simply pushing their way through the floor of the ocean. Other scientists recognized this explanation to be very unlikely and yet could not offer a satisfactory alternative, so the question remained: How, or by what *mechanism*, do the continents move around the earth?

REMEMBER

After many decades of research, scientists currently accept the idea that movement of heated materials in the mantle of the earth, or *mantle convection*, plays a large role in driving the continental plates around the surface of the earth. I explain the details of mantle convection and plate movement in Chapter 10.

Coming Together: How Technology Sheds Light on Plate Tectonics

Since the early part of the twentieth century, there have been major breakthroughs in understanding the earth's continental plates. *Marine geology* (the study of geology and geologic features in the oceans) and advances in military technology have led the way.

Mapping the seafloor

During World War I, the introduction of submarines in warfare meant that the military needed extensive, detailed maps of the seafloor. When the seafloor of the Atlantic Ocean was mapped, it revealed an interesting feature: a chain of undersea mountains running north–south with a deep crack, or valley in the middle of the mountain chain. This crack, or *rift*, scientists now know, is the boundary along which new ocean crust is created.

IMAGING THE SEAFLOOR: MARIE THARP AND SEAFLOOR MAPPING

Mapping the seafloor is not as simple as it sounds. It's not as if we can simply remove the water and take pictures of what is below. Rather, ships must cross back and forth while using sonar, or sound waves, to capture the highs and lows, or *bathymetry*, of the seabed. Each crossing creates a profile using the sonar data that must then be translated into features such as mountains and valleys on a map. The seafloor map scientists use today is the result of work by two mid-twentieth century geologists, Marie Tharp and Bruce Heezen. Marie Tharp used the collected sonar bathymetry data and artistically interpreted it into a hand-drawn map of features of the seafloor. As Tharp worked on her map, she and Heezen recognized the Mid-Atlantic Ridge feature and suspected that it might be the final piece of the puzzle of plate tectonics — the place where tectonic plates are pushed apart. To learn more about Marie Tharp and her important work, check out the book *Soundings: the Story of the Remarkable Woman Who Mapped the Seafloor* by Hali Felt.

Mapping the seafloor during the early twentieth century led to other studies, including those investigating the age of the rocks beneath the ocean. The results of dating seafloor rocks finally provided the key insight scientists were looking for to explain how the continents moved around the surface of the earth. Had it not been for submarine warfare, funding for undersea mapping may not have been available, and who knows when scientists would have uncovered the key bit of information they needed to draw together, or unify, the fields of earth science.

Flip-flopping magnetic poles: Paleomagnetism and seafloor spreading

To explain how scientists date rocks on the seafloor, I first need to describe an important characteristic of the earth: magnetism. You are familiar with the earth's magnetism if you have ever used a compass. The needle in a compass points to magnetic north (or south if you are in the southern hemisphere). Similarly, when magnetic minerals such as magnetite (formed of iron) form, they point, or align, towards the magnetic poles. This observation by scientists has led them to understand that earth's magnetic poles do not remain in the same position all of the time. Sometimes the position of the magnetic poles (which right now are very close to the geographic poles) shifts slightly. The movement of the magnetic poles is called *polar wandering*. Before geologists understood tectonic plate movements, they assumed that the magnetic rock record on the continents recorded dramatic and inconsistent shifts in the location of the poles. This is referred to as *apparent polar wandering*. However, when it became accepted that the continents moved long distances across Earth's surface and the poles didn't shift dramatically, scientists considered the idea of apparent polar wander *debunked*, or proven incorrect.

Earth's magnetic poles also occasionally switch so that the magnetism recorded from the current geographic North Pole comes from the geographic South Pole instead. This phenomenon is called a *magnetic reversal*.

REMEMBER

Paleomagnetism is a record of how the earth's magnetic poles have changed direction over time. Minerals in rocks are aligned toward the poles when the rock is formed and record a history of these changes. Rocks formed with their minerals aligned to indicate the same magnetism as the present have *normal polarity*, while rocks created during a reversal period, with minerals aligned in the opposite direction, have *reversed polarity*.

These patterns of polarity are recorded in the minerals of basalt rocks (see Chapter 7) created along the Mid-Atlantic Ridge of the seafloor (and in basalt formed during volcanic eruptions all over the world). By combining all this data with other methods of geologic dating (explained in Chapter 16), scientists can create a magnetic time scale for the earth.

SHARKS AND THE EARTH'S MAGNETISM

Researchers have discovered that sharks use the earth's magnetism to help guide them through the oceans. In the front of their heads, sharks have small electromagnetic sensors that detect the strong north-south magnetic currents of the earth. These same sensors also pick up the much weaker electromagnetic currents emitted by other animals. For example, the great white shark can orient itself toward mammalian prey several feet away by detecting the electromagnetic pulses of its heartbeat traveling through the water.

Scientists have mapped parallel patterns of magnetic reversals along each side of the Mid-Atlantic Ridge. Dates for the rocks indicate that rocks collected from the ridge itself are younger than rocks located farther from the ridge. The relative ages of the seafloor rocks are illustrated in Figure 8-7.

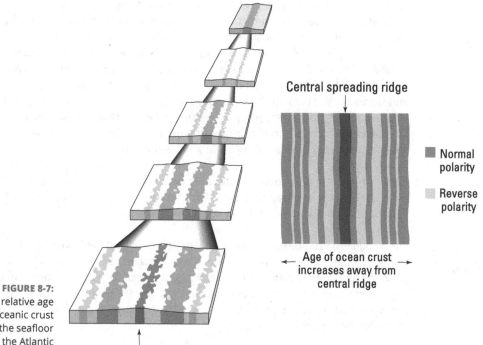

FIGURE 8-7:
The relative age of oceanic crust along the seafloor of the Atlantic Ocean.

Central spreading ridge

■ Normal polarity

■ Reverse polarity

← Age of ocean crust → increases away from central ridge

Central spreading ridge

The sequence of magnetic reversals and patterns of older rocks situated farther from the rift indicate that the oceanic plates are spreading out from the Mid-Atlantic Ridge as new rock is formed. This evidence leads to the conclusion that new ocean crust is being created along the ridge as the oceanic plates move apart from one another. (I further explain the creation of rocks along the Mid-Atlantic Ridge and in other similar areas of the earth in Chapter 10.)

Measuring plate movements

After scientists recorded that crustal plates had indeed moved, they sought a way to determine the rate, or *velocity*, of plate movement.

TECHNICAL STUFF

They used their record of the seafloor rock ages and the distance of the rocks from the Mid-Atlantic Ridge to calculate the distance each plate moved per year. The answer is approximately 18 millimeters per year for the last 3 million years.

As they continued to map the magnetism and ages of the seafloor rocks, scientists noticed that the rate of plate movement varies quite a bit. Along the Mid-Atlantic Ridge, the crust appears to be spreading more slowly near Iceland than it is near more southern parts of the ridge.

To refine calculations of plate velocities, modern scientists have begun using the Global Positioning System (GPS) of satellites in orbit around the earth. You probably have a GPS system in your car or phone that is very similar to (though much less precise than) the one scientists use to measure plate movements. The changes in distance and the direction of plate movement have been recorded by the GPS satellites. These findings confirm previous calculations based on magnetic reversals in the seafloor basalts. Now geologists rely on GPS tracking of the plates to help answer questions about current plate movements and tectonic processes, which I describe in Chapter 9.

Unifying the theory

What began with ideas about continental drift has been developed into a foundational explanatory theory of how Earth's crustal plate system functions now and functioned in the past. The confirmation of seafloor spreading in the Atlantic Ocean was just the first step.

If the seafloor is spreading along the Mid-Atlantic Ridge, what is happening along other edges of the crustal plate? This is where plate tectonics theory extends its explanation to include mountain building, volcanoes, and earthquakes, which I explain in Chapter 9.

IN THIS CHAPTER

» **Understanding density**

» **Recognizing two types of crust**

» **Examining the role of density in plate interactions**

» **Defining plate boundaries, relatively**

» **Deforming rocks and building mountains**

Chapter **9**

When Crustal Plates Meet, It's All Relative

With the theory of plate tectonics (see Chapter 8), geologists finally had a single explanation that could describe earth processes all around the globe. The relationships among features that seem so different, such as mountains, earthquakes, and deep sea trenches, suddenly made a little more sense.

The key to understanding how geologic features and earth processes are related is understanding crustal plate movement. The movement of multiple, individual plates of earth's crust creates and destroys many of the geologic features you are familiar with.

In this chapter, I describe the different types of crust found on the earth's surface, including their different compositions and characteristics. I also explain how the many separate plates that cover the earth move around and what happens when they collide or move apart. These motions and the relationship between plates create mountains, earthquakes, ocean trenches, and volcanoes. Finally, one theory to explain them all!

Density Is Key

In order to understand the interactions between crustal plates described in this chapter, you must first conquer the concept of *density*.

The density of an object is described as the mass per unit of volume. *Mass* is a measure of the amount of matter in an object (usually measured on a scale similar to weight). *Volume* is how much space the object fills. A formula describing this relationship is written:

$$D = m/v$$

In this formula,

>> m = mass (measured as grams, or g)

>> v = volume (measured as cubic centimeters, or cm^3)

>> D = density (expressed as grams per cubic centimeter, or g/cm^3)

To put it another way, density is the amount of stuff (matter) in a set volume of a certain material. While density is *related to* the size (the volume), it is not *determined by* the size (or volume). For example, two objects of the same size (or volume) may have different densities. The difference is a result of one of the objects having more mass filling that same space.

Consider two identical kitchen sponges. One is dry, and the other is saturated (filled) with water. They continue to take up the same space, having the same volume, but the dry sponge feels lighter in weight — it is less dense. In the saturated sponge, water fills the holes of the sponge, adding more matter, increasing the mass, and therefore increasing the density.

REMEMBER

Mass is often confused with weight because you measure mass and weight in similar terms, using a scale. But there is an important difference: Weight measures the pull of Earth's gravity on an object, while mass measures the amount of atomic matter in the object (see Chapter 5 for details on atoms). For example, if you go to the moon, you will weigh less because the moon's gravity does not pull as strongly as Earth's gravity. But even on the moon, you will have the same mass as you have on Earth — the same amount of atomic matter composing your body.

Two of a Kind: Continental and Oceanic Crust

Covering the earth is the *lithosphere*, a layer of crustal rock that I describe in Chapter 4. All the rocks in the lithosphere are made of basically the same stuff: silicate minerals (see Chapter 6). However, there is a slight difference between the minerals found in crustal rocks under the oceans and those found in crustal rocks that make up the continents. In this section, I describe the differences between these two types of crust and discuss a little about *why* they are different.

Dark and dense: Oceanic crust

The lithosphere beneath the oceans is called *oceanic crust.* Oceanic crust is created from molten mantle rock (magma) rich in dark-colored, dense minerals. The molten rock erupts along ridges in the ocean (*mid-ocean ridges,* which I describe later in this chapter and in Chapter 10) and cools, forming basalt and gabbro rocks (see Chapter 7). Because of the types of minerals that comprise these rocks (such as olivine and pyroxene; see Chapter 7), the oceanic crust is dark in color and relatively dense compared to continental crust. The average density of oceanic crust is 3.0 g/cm³. Oceanic crust is also relatively thin and about the same thickness all around the world — about 8 kilometers (5 miles).

Thick and fluffy: Continental crust

The continents are composed of *continental crust.* Continental crust is made of the less dense minerals typical of granite-type rocks. Typically, magma that produces these less dense minerals experiences partial melting and fractional crystallization before erupting at the surface or cooling as a pluton beneath the surface (see Chapter 7). The thickness of continental crust varies widely. In some areas, it is relatively thin, about 25 kilometers (15.5 miles) But in mountainous regions it may be up to 70 kilometers (43.5 miles) thick. Its average density is 2.7 g/cm³.

Table 9-1 summarizes the differences between oceanic and continental lithosphere.

TABLE 9-1   **Characteristics of Oceanic and Continental Crust**

Crustal Type	Average Density	Average Thickness	Common Minerals
Oceanic	3.0 g/cm³	7–10 km	Olivine, pyroxene
Continental	2.7 g/cm³	25–70 km	Quartz, feldspar

Understanding Why Density Matters: Isostasy

The different densities of oceanic and continental crust affect the way each one interacts with the mantle material below it. The mantle material beneath the earth's crust is solid rock, but it moves a little via plastic flow, which I describe in Chapter 13. The result of this semi-fluid characteristic of the mantle rock is that the solid lithosphere above sinks into it slightly.

REMEMBER

The density of a floating object relative to the material it floats in determines how much of the object sinks into, or displaces, the liquid. Because density is the relationship between mass and volume, the point at which the volume of displaced liquid has the same mass as the object that is displacing it is marked by the *equilibrium line*. This is illustrated in Figure 9-1.

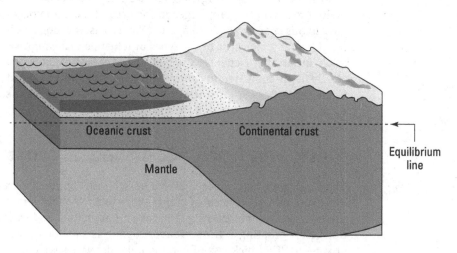

FIGURE 9-1:
The equilibrium line for continental and oceanic crust.

Therefore, the density of the crustal plate determines how deeply it sinks into the mantle rock. The more dense oceanic crust sinks more of its volume into the mantle than the less dense continental crust.

However, continental crust is usually much thicker than oceanic crust. Figure 9-1 illustrates oceanic crust and continental crust above earth's mantle. Because the continental crust is so much thicker than the ocean crust, it may appear that much more of it is submerged into the mantle materials. But if you compare the relative proportions or percentages of submerged continental and oceanic crust, you see that a larger percentage or proportion of the oceanic crust sinks into the mantle rock. Only 80 percent of continental crust is submerged into the mantle. (It only appears to be more on first glance because the continental crust is so thick.)

A larger percentage (a little more than 90 percent) of the oceanic crust is submerged as a result of its higher density.

TIP

One way to think about the relationship between the mantle and the continental crust is to think about an iceberg floating in the ocean. Part of the iceberg is above the water surface, but a larger percentage of it is below the water where you can't see it. The portion of the iceberg above and below the water line is determined by the density of the ice. In the case of the iceberg, approximately 10 percent of the overall volume of the iceberg is above water, and 90 percent is below. This is because the iceberg is only slightly less dense (10 percent) than the water it is floating in.

The position of equilibrium — where the density of the portion of crust sunk into the mantle is equal to the density of the mantle material displaced — is called *isostatic equilibrium.* Any time material is moved around the earth's surface, such as when rocks and sediments are removed from a mountain by erosion, the crust adjusts its position in the mantle; it finds a new balance between the portion of crust above and below the line of equilibrium in order to accommodate the material removed in one place and added in another. This balance of the solid crust floating in the solidly flowing mantle is called *isostasy.*

Understanding the different densities and thicknesses of each type of crust, as well as how the two types of crust sit in the mantle differently, helps you understand what happens when two plates meet, which I explain in the following sections.

Defining Plate Boundaries by Their Relative Motion

The crustal plates don't move in a single, specific direction, such as north, south, east, or west. They just move. Some, like the North American plate, seem to rotate in a kind of circular motion where portions of the plate are moving towards nearby plates, whereas in others portions are moving away from nearby plates. Because of the complicated motion of the tectonic plates, when describing how the crustal plates move around the surface of the earth, the best way to define their motions is in relation to one another at specific locations. The features that occur at locations where plates interact are the clues to determining whether the plates are coming together, moving apart, or slip-sliding past one another.

The relative motion of plate boundaries can be narrowed down to three categories:

>> **Divergent plate boundary:** The boundary between two crustal plates that are moving apart from each other

>> **Convergent plate boundary:** The edge along which two crustal plates meet when they are moving toward one another

>> **Transform plate boundary:** The boundary between two crustal plates that are neither moving toward each other nor apart — they are sliding back and forth next to each other

WARNING

Be careful not to confuse the edge of a continent with the edge of a crustal plate. In some cases, such as along the west coast of South America, the two edges are the same. When the edge of the continent is also the edge of the plate, it's called an *active continental margin*. It is active because the plate boundary is engaged in geologic processes (such as earthquakes and volcanoes) as a result of its interaction with another crustal plate.

But in some cases, like along the eastern coast of North America, the crustal plate extends into the ocean and includes continental lithosphere as well as oceanic lithosphere. In this case, the edge of the continental portion of the plate is called a *passive continental margin* because it is not actively participating in plate boundary processes.

In the following sections, I explain what happens to the crustal plates and what features are created at the earth's surface at each of these boundary types.

Driving apart: Divergent plate boundaries

When two crustal plates move apart from each other, the edge along which they are separating is a divergent boundary. The particular characteristics of a divergent boundary vary depending on whether the two plates are continental crust or oceanic crust.

Ripping open the ocean floor

In many places around the earth, two plates are moving apart from each other. This is most common on the sea floor, where this separation allows molten mantle rock (magma) to erupt along the boundary, harden, and create new ocean floor. (I describe this process in detail in Chapter 10.) The buildup of this new rock material creates a *mid-ocean ridge* along the undersea divergent boundary. Mid-ocean ridges form the longest continuous mountain chains on earth, winding through all the oceans.

Many of the mid-ocean ridges also have a rift valley. A *rift valley* is created as the two plates pull apart and the freshly created, cooling rocks between them are stretched until they *fault* (break) and drop down, creating a valley. This process is illustrated in Figure 9-2.

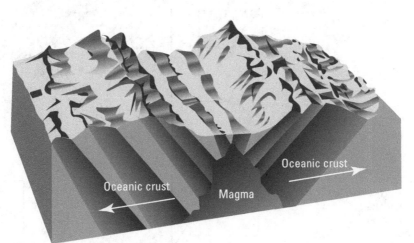

FIGURE 9-2:
Characteristics of a mid-ocean ridge and rift valley.

The seafloor spreading that occurs along oceanic divergent boundaries, or mid-ocean ridges, creates new seafloor and builds oceanic crust of basalt. (I describe this process in Chapter 10 as well.)

Parting the Red Sea

In some areas of the world, divergent boundaries occur where continental crust is spreading apart instead of oceanic crust. It is common for continental divergent boundaries to begin as a *triple junction*, where the crust splits apart in three directions.

In most triple junctions, evidence shows that only two of the cracks continue to separate. For example, in Figure 9-3 the triple junction between the Arabian Plate and two parts of the African Plate is illustrated.

In this region, the Red Sea and the Gulf of Aden are actively spreading, while Africa's rift valley has stopped spreading. Africa's rift valley is called the *failed rift*, while the Red Sea and Gulf of Aden are *active rifts*.

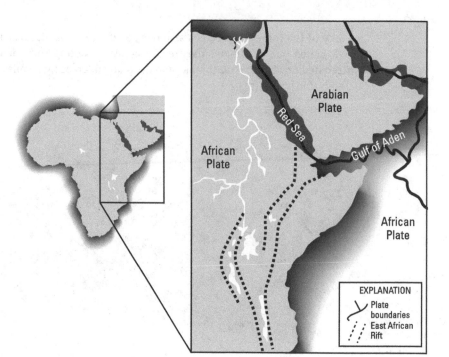

Rifting of continental crust begins when heated mantle rock rises up underneath the continent, forcing the crust upward until the stress causes it to crack. As the continental crust moves apart, eventually a rift is wide enough that nearby ocean water spills in, creating a small sea. For example, the rift along the floor of the Red Sea began as a rift between the continental rocks of Egypt and the Arabian Peninsula. Already the ridge along the floor of the Red Sea is creating new basalt rock. As rifting continues, this small sea becomes the first stage of a new ocean basin.

REMEMBER

This means that what starts as a continental divergence eventually becomes an oceanic plate divergence, when the continental plates are far enough apart that ocean lies between them. Scientists think this series of events is how all ocean basins begin.

Crashing together: Convergent plate boundaries

When two crustal plates move toward each other, they create a convergent plate boundary. At a convergent boundary, part of the lithosphere is forced downward into the mantle while other parts of it may be forced upward, building mountains. Exactly how this process occurs, and to what degree, depends on the type (and therefore density) of the crustal plates.

Three convergent boundary types are possible, as I describe in this section.

One goes up and one must go down

The most common convergent plate boundary, and the easiest to recognize, is one where a plate of continental crust meets a plate of oceanic crust. When this occurs, the more dense oceanic crustal plate is forced downward beneath the continental crustal plate in a process called *subduction*. Regions where this occurs are called *subduction zones*. Most subduction zones share certain features, such as a trench on the seafloor where one plate disappears beneath the other, or a nearby chain of volcanoes and earthquakes in a pattern of gradually deepening locations, as illustrated in Figure 9-4.

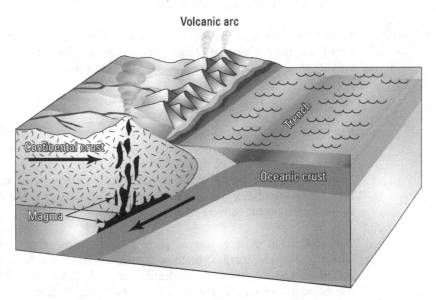

Volcanic arc

Continental crust

Trench

Oceanic crust

Magma

FIGURE 9-4: A continental-oceanic plate convergent boundary subduction zone and associated geologic features.

As the dense oceanic plate is forced downward, or *subducted*, it rubs against the continental plate, causing friction that creates heat. The heating releases carbon dioxide and water stored in the oceanic crust, which enters the crustal rock above the subducting plate. This addition of volatiles causes flux melting, and the causes rocks in the lithosphere above the subducting plate to become a liquid (magma). (For more details on how volatiles cause rock to melt, visit Chapter 7.) The density of the rock materials decreases dramatically after they are melted. The molten rock material then begins to move upward through the continental crustal rocks of the overlying plate until it erupts onto the surface.

REMEMBER

As all the magma from beneath the continental crustal plate breaks through the surface, a chain of volcanoes is created along the surface, parallel to the edge of the subducting oceanic plate. This feature is called a *continental volcanic arc*. The Cascade Mountain range along the West Coast of the United States is an example of a volcanic arc created by subduction of an oceanic plate under a continental plate.

A subducting plate must reach a certain depth before it is hot enough to release volatiles and cause melting. As the plate moves downward away from the trench where it first dives beneath the surface, it moves at an angle beneath the overlying plate. When it reaches the depth where it melts, the resulting magma moves directly upward to produce volcanic features at the surface. The distance from the trench to these volcanic features (usually a line of volcanoes called an *arc*) indicates to scientists the angle of its movement into the mantle. If the trench and the volcanic arc are close together, the *subduction angle* is steep. This means the plate reaches the right depth for melting more quickly on its journey downward. On the other hand, if the trench and the volcanic arc are far apart on the earth's surface, then the subduction angle is less steep.

REMEMBER

The friction of the plates against one another also causes earthquakes to occur near the surface where the plates meet; gradually they occur more deeply as the oceanic plate is subducted beneath the continental plate.

Diving into the abyss

When two plates of oceanic crust move toward each other, they create an *oceanic convergent boundary*. The force of their movement causes the older, colder, and slightly denser plate to be driven beneath the other, down into the mantle rocks below. This interaction of oceanic plates is another type of subduction zone, and the resulting features are in some ways similar to those I describe for a continental-oceanic convergence.

The subduction of one oceanic plate beneath another oceanic plate creates friction, releases volatiles, melting the lithospheric mantle rock and creating volcanoes. But in this case, the volcanoes erupt beneath the ocean rather than on a continent. Eventually, the volcanoes along the seafloor may build up high enough to be visible above the surface of the sea, creating a *volcanic island arc*. The islands of Japan are an example of a volcanic island arc.

REMEMBER

A unique feature occurs along the subduction boundary of two oceanic plates, due to their similar density. As the subducting plate moves downward into the mantle, it pulls the overlying plate downward slightly, creating a deep sea *trench*. This feature and other oceanic-oceanic plate subduction zone characteristics are illustrated in Figure 9-5.

Reaching for the sky

A third type of convergent boundary occurs when two plates of continental crust crash into each other. This is called a *continental-continental convergent boundary*.

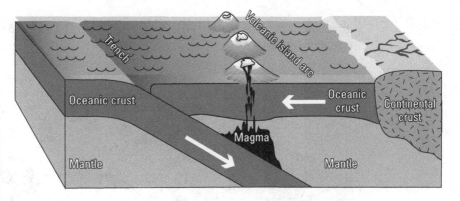

FIGURE 9-5:
An oceanic-
oceanic
convergent plate
boundary and
associated
geologic features.

REMEMBER

When two plates of continental crust come together, they are each composed of similar, relatively low density crustal rocks so neither sinks and subducts as easily as the oceanic crust of other convergent boundaries. Instead, rocks from each continental plate pile up onto each other, building up tall mountains as the plates continue to push together. Some crustal material will be forced into the mantle below, but because of the lighter density of the continental lithosphere, the crustal rocks resist subduction, resulting in a buildup of crustal material creating mountains. The features of a continental-continental plate convergence are illustrated in Figure 9-6.

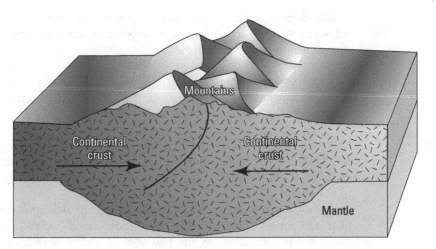

FIGURE 9-6:
A continental-
continental plate
convergent
boundary and
associated
geologic features.

Slip-sliding along: Transform plate boundaries

Sometimes, where the edges of two plates meet, they are neither moving toward one another nor away from one another. Instead, they are just sliding in opposite directions. This type of plate boundary is called a *transform boundary* and is illustrated in Figure 9-7.

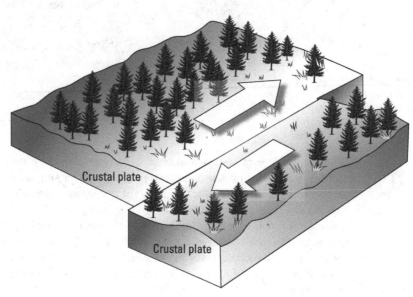

FIGURE 9-7:
Features of a
transform
boundary.

At transform boundaries, the interaction between the two plates is much more subtle; there is no direct pushing, no subduction, no stretching of the plates, and no production of new rocks from magma.

REMEMBER

Earthquakes are common at transform boundaries as the plates grind, slip, and slide past one another. The earthquakes occur very close to the surface (because there is no subduction) and can be very powerful.

An example of a transform boundary is the San Andreas Fault in California, pictured in this book's color photo section. Along this fault, shallow earthquakes are common as the North American Plate slides past the Pacific Plate.

Transform boundaries, or *transform faults*, also occur along mid-ocean ridge divergent boundaries. These transform boundaries associated with mid-ocean ridges are called *fracture zones*. As the two oceanic plates move apart, stretching the newly created crustal rocks, some of the material responds to the pull by breaking and slipping to one side or the other. This process creates a zig-zag pattern of transform boundaries all along the divergent ridge, as illustrated in Figure 9-8.

Transform boundaries transform rocks primarily through shear stress, which I describe in the next section.

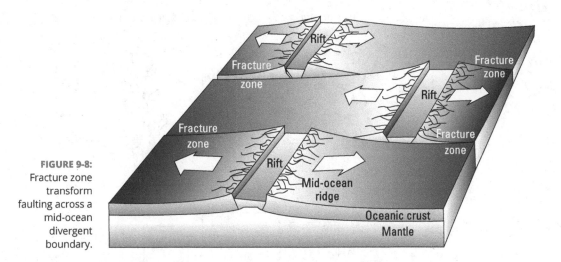

FIGURE 9-8:
Fracture zone transform faulting across a mid-ocean divergent boundary.

Shaping Topography with Plate Movements

The movement of crustal plates around Earth's surface creates immense amounts of pressure and strain on the solid rocks that comprise the lithosphere. These forces stretch, bend, snap, fold, fracture, and crumple the thick, solid layers of rock on Earth's surface. In response to these kinds of physical stress, rocks *deform* or change shape.

In this section, I describe the different ways rock is deformed at plate boundaries and how the shifting of crustal plates, combined with deformation, results in the extreme topography of mountains.

Deforming the crust at plate boundaries

Because the relative motion at the different plate boundaries (convergent, divergent, transform) is different, the rocks at each plate boundary type are deformed in different ways.

Compression is what occurs at convergent boundaries where two plates move toward one another and compress or crush the rocks in between them. The opposite occurs at divergent boundaries, where the plates are moving apart and create *tension stress* by stretching the rocks between them. And at transform boundaries, as the plates move in opposite directions alongside (parallel to) one another, the rocks between them experience *shearing stress* as they break apart and move with one plate or the other. Each type of stress is illustrated in Figure 9-9.

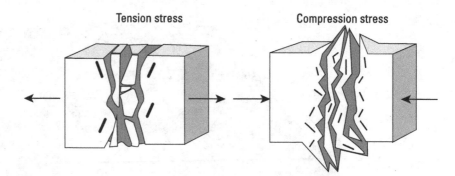

Tension stress   Compression stress

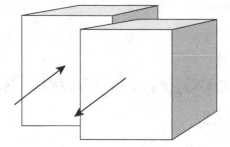

Shear stress

FIGURE 9-9:
Three types of
rock stress.

When an object responds to stress by simply breaking, it experiences *brittle failure* or *brittle deformation*. This response may happen with rocks if the stress is applied suddenly, especially if the rocks are near Earth's surface and relatively cool.

Deeper in the earth's crust, rocks are exposed to higher heat and slowly building pressure. These rocks are more likely to respond to stress with *ductile* or *plastic deformation*: a change in their shape without breaking or fracturing.

Compressing rocks into folds

At convergent plate boundaries, rock layers are often compressed into folds. Folds in rocks may be small enough to see in a rock you hold in your hand or large enough to create entire landscapes, such as the Rocky Mountains of western Canada. When rocks are folded, they crinkle together similar to bunched-up or gathered fabric.

Folds often occur deep in the crust where the rock layers are exposed to high temperatures and pressures. When these rocks are compressed by convergent plate movements, they respond plastically by crumpling together or folding.

Scientists categorize folded rocks with different terms that describe the nature of the folded geography:

» **Monocline:** *Monoclines* are the simplest feature of folded rocks, with layers that are bent only a little.

» **Anticline:** *Anticlines* are folded rock layers that create an arch (see Figure 9-10).

» **Syncline:** *Synclines* are the opposite of anticlines. Synclines are the U-shaped folds found between anticline folds (see Figure 9-10).

» **Dome:** *Domes* are rounded or oval-shaped areas of rock that are lifted slightly in the center (bulging), creating the appearance of an anticline. After domes are eroded, they are recognized by the older rocks in the center and younger rocks around the outside (see Figure 9-11).

» **Basin:** Basins are similar to domes, but instead of bulging, they dip downward in the center and create a bowl-shape depression that appears like a syncline. An eroded basin is recognized by the younger rock layers in its center and older rock layers around the outside (see Figure 9-11).

In regions that experienced folding, you will find foliated metamorphic rocks (see Chapter 7) that indicate the degree of pressure and temperature change that occurred as a result of compression.

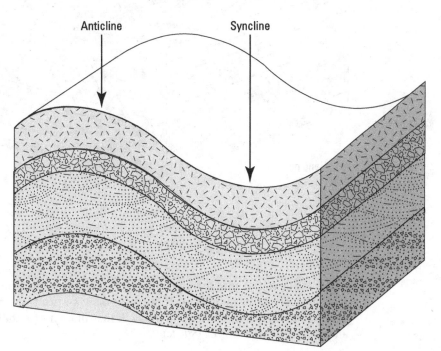

FIGURE 9-10: Anticline and syncline features.

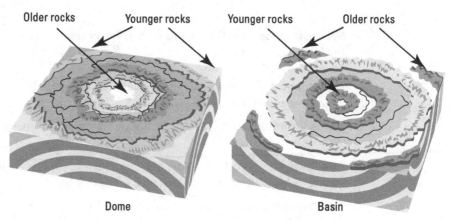

Older rocks   Younger rocks   Younger rocks   Older rocks

FIGURE 9-11:
Dome and
basin features.

Dome   Basin

Faulting in response to stress

Closer to the surface, the rocks are cooler and more brittle. When these rocks respond to stress, they are more likely to *fault*, which means fracture and crack. The two pieces of rock that separate, creating a fault, are called the *fault blocks*; they separate along the fault plane, as illustrated in Figure 9-12.

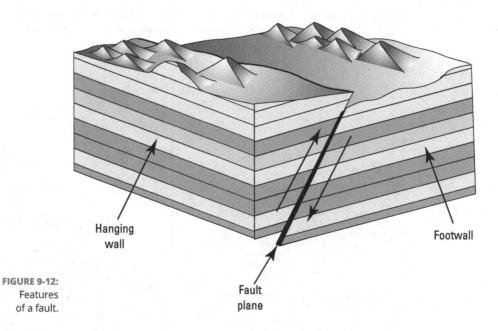

Hanging
wall   Footwall

FIGURE 9-12:
Features
of a fault.   Fault
                               plane

Faults are classified into two major categories based on the primary direction of movement along the fault plane. If the rocks move vertically (up or down) relative to one another, they create a *dip-slip fault*. If they move horizontally relative to one another, they create a *strike-slip fault*.

Dipping and slipping

Dip-slip faults are the result of rocks breaking apart and moving up or down next to each other along a tilted fault plane. This movement creates two separate blocks of rock: a *hanging wall* above the fault plane and a *footwall* below the fault plane.

TECHNICAL STUFF

The names *hanging wall* and *footwall* for the blocks of a dip-slip fault come from mining terminology. Important minerals and metal ores are often concentrated in veins that follow fault fractures. When miners would set up to work removing these materials from the vein, they hung their lanterns on the rocks above the vein, or fault (thus, the hanging wall), and they stood at the rocks below the fault (thus, the footwall).

A *normal fault* is a dip-slip fault where the hanging wall block slips downward in relation to the footwall; this is a response to tension stress that allows the rock to break, and drop downward, as illustrated in Figure 9-13. Normal faults like this are common at mid-ocean divergent boundaries, where the outward motion of the plates fractures the rocks through tension stress, while creating space where the rocks can slip downward.

A *reverse fault* is the result of compression, where the hanging wall moves upward in relation to the footwall, accommodating the movement of two plates toward one another at a convergent boundary. When the angle of the fault plane is low, the result is a *thrust fault,* where the hanging wall is thrust up and slightly over the footwall rocks. A reverse fault is also illustrated in Figure 9-13.

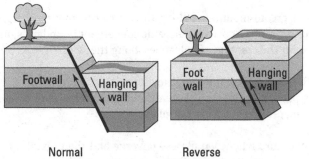

FIGURE 9-13:
Dip-slip faults.

Normal

Reverse
(or thrust)

Striking and slipping

Strike-slip faults occur most often along transform boundaries as a result of shearing stress. In a strike-slip fault, the rocks move horizontally, side by side. Strike-slip faults are often linear (like a line) and can be miles long across the surface. For example, the San Andreas Fault (pictured in this book's color photo section) is a 600-mile-long strike-slip fault. Rather than a single long fracture, long strike-slip faults are usually a series of parallel fractures running in the same direction.

Smaller strike-slip faults are found along mid-ocean ridge divergent boundaries. As the crustal plates move apart, blocks of rock slip in one direction or the other and create a zig-zag pattern of transform faults, called the *fracture zone,* as illustrated earlier in this chapter in Figure 9-8.

Joints

In some cases, fractures or faults appear in rocks where there does not appear to have been any movement. These fractures are called *joints.* Joints are common at the surface of crustal rocks. While the lower layers of rock are folded, the rocks at the very surface, in the bend of the fold, may fracture apart but not move in relation to one another.

Building mountains

REMEMBER

Geologists call the process of mountain building *orogenesis.* Orogenic processes are a result of plate movements and the various processes observed at the different plate boundaries. Some mountains begin as volcanoes; others are the result of crustal plates being stretched apart or crushed together.

Volcanoes and accretion

Volcanoes create mountains along the active continental margin of a subduction zone, for example. At oceanic-oceanic convergent boundaries, mountain-building begins with the creation of volcanoes along the seafloor.

As seafloor volcanoes build up, they form island mountains. Over time, with continuous plate movements these volcanic mountains may be *accreted* or added to the continental crust of another plate.

REMEMBER

Accretion occurs when two plates converge and the crustal materials of the subducting plate are attached to the edge of the overlying plate. This process is illustrated in Figure 9-14.

The accumulation of sediments and rocks at a subduction trench is called an *accretionary wedge.* Often these rocks are later uplifted and added to the edge of the nearby continental plate. The Olympic Mountains in Washington State are an example of an accreted wedge.

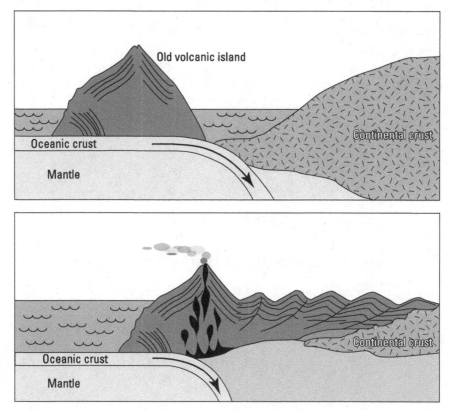

FIGURE 9-14: Accretion of volcanic islands onto continental crust.

Stretching and thinning

In regions where the crust is uplifted by hot magma, such as along the mid-ocean divergent boundaries, the tension stress and resulting faults create *fault-block mountains.* An example of fault-block mountains formed long ago is the basin and range province of Nevada in North America. In this region, the crust has been lifted and stretched, tearing the rocks apart in such a way that large blocks of rock called *grabens* drop down between raised blocks of rock called *horsts,* creating a series of parallel mountain ridges.

Crushing and lifting

More common than fault-block mountains are *fold-thrust mountains,* which occur at convergent boundaries where two plates of continental crust crash together. The compression force of the two plates moving toward each other folds the rock layers, deforming them and thrusting them upward, creating mountains such as the Himalayas or the Alps. In fact, this orogenic process occurs today as India is pushed farther up toward Asia, continuing to uplift and build the Himalayan mountain system.

Chapter **10**

Who's Driving This Thing? Mantle Convection and Plate Movement

O ne of the biggest challenges for Alfred Wegener when he proposed his idea that the continents had once been connected and were now moving apart (see Chapter 8) was describing a *mechanism*, or driving force, for continental movements. He supposed that the continents just plowed through the ocean crust — like a whaling ship through sea ice — as they careened around the globe crashing into one another. Today, we understand that the continents above sea level are part of larger plates made of both continental crust and oceanic crust (see Chapter 9), but how and why they move around is still somewhat of a mystery.

Earth scientists now accept that the movement of crustal plates is very likely related to *convection* — the movements of heated materials — inside the earth. But they have not agreed on *how* convection drives the plates. The old understanding of mantle convection to move plates relied on a simple model of convection cells moving in a circular motion, and assumed the plates simply rode along the top of these cells. However, with modern imaging of the mantle rock, scientists now understand that the story is not so simple. Although heated materials in the mantle do move upward, other forces acting on plate movement include the slab-pull force of the sinking plate and the ridge-push force of the elevated mid-ocean ridges. In this chapter, you explore these different ideas. I also explain how mantle motion and plate movements create volcanoes. And finally, I explain how the movement of plates causes earthquakes and how studying earthquakes provides additional clues about the earth's interior.

Running in Circles: Models of Mantle Convection

Current models describing plate movement rely on *convection* as their driving force in one way or another. Convection is, quite simply, the movement of heated materials. You probably know it as the concept that "heat rises." The trouble with that phrase is that *heat* itself doesn't actually go anywhere. Instead, the *material* that is heated — whether it is solid, liquid, or gas — starts moving.

When matter is heated, the molecules start moving due to the increase in energy. As they get moving, they occupy more space; they spread out and become less dense. The number of molecules stays the same, but they take up more space or *volume.*

Think of a dance floor: During a slow song, all the couples embrace and sway slowly together, and you can really pack the floor with couples! Then the deejay switches to a dance tune, and people start to boogie. Suddenly, with all that energy and movement, people are taking up more space, and some of them have to move off the dance floor and dance between the tables because there just isn't enough room anymore. You still have the same number of people, but the dancing crowd becomes less dense and takes up more volume because you added some energy.

REMEMBER

In a similar way, the rock material in the *mantle* (the layer between earth's crust and core, as described in Chapter 4) is heated and begins to move. When the mantle rock is near the core, it heats up, becoming less dense. This heated mantle rock then moves up, toward the earth's crust and away from the core, forcing the cooler and more dense mantle rock near the crust down toward the core. This circular rotation of material creates what's called a *convection cell.*

Some scientists think that only the upper part of the earth's mantle moves through convection cells; other scientists think that the entire mantle rotates this way. By observing the change in speed of seismic waves that move through the mantle, scientists have seen that hotter and cooler regions of the mantle do exist, but they do not appear to move in simple, circular, convection cell-like motion. (See Figure 10-1. More details on how scientists use seismic waves to understand earth's interior can be found in Chapter 4.)

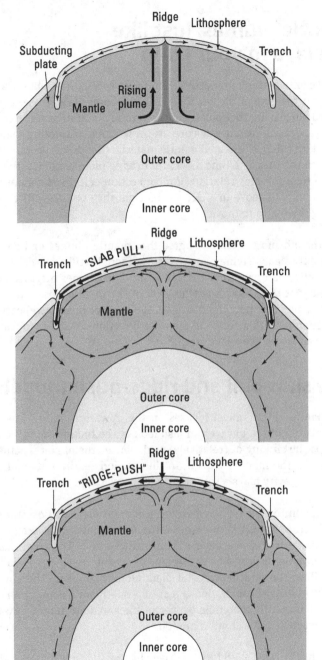

FIGURE 10-1:
A cross-section of the earth illustrating the focus of each model of mantle convection: mantle plumes rising from the earth's core; slab-pull by plates into the mantle; and ridge-push of plates outward and down into the mantle.

Mantle plumes: Just like the lava in your lamp

Some scientists think the earth's mantle works very much like a lava lamp. Mantle rock near the core of the earth is heated, becomes less dense, and begins to flow upward, displacing the cooler, denser mantle material near the crust, which sinks downward, back toward the core. In these loops of rotating mantle material (or convection cells), the heated mantle material that drives upward is called a *mantle plume.* In the mantle plume model for driving plate tectonics, the rising of heated mantle rock initiates rotation of a convection cell. As the mantle plume–driven convection cells move in a circular motion, they pull the crust along with them, causing the plates to move.

Even though most scientists agree that mantle plumes and convection play an important role in driving plate tectonics, there is still some debate on exactly how the mantle material circulates and where the plumes originate. Some scientists propose that the *mesosphere* (the layer of the mantle closest to the core) and the *asthenosphere* (the part right beneath the crust or *lithosphere*) each have a layer of convection cells that rotate in opposite directions. Others propose that the entire mantle convects like a lava lamp.

The slab-pull and ridge-push models

The mantle plume model focuses on the upward motion of heated material to drive plate motion, but we can also look at the motion from the other direction. Perhaps the sinking of (relatively) cold, dense crustal plates actually drives plate motion. In the model I explain in this section — the slab-pull and ridge-push model — gravity plays a role along with convection.

As I explain in Chapter 9, when a continental plate and oceanic plate meet, the oceanic crust (which is more dense) is forced down *(subducted)* into the mantle beneath the less dense plate of continental crust. The term *slab-pull* describes plate movements driven by the sinking of oceanic crust (the "slab") into the mantle, pulling the attached crustal plate along behind it. The sinking slab is the downward arm of cooler, dense material in a mantle convection cell. As it sinks, the heated mantle materials deeper in the earth are forced upward, completing the convection cell's circular motion.

Similarly, the *ridge-push* force is driven by gravity. In this model, the heated mantle below a mid-ocean ridge (a plate boundary where new oceanic crust is being created) slightly lifts the crustal plates as it wells up, erupting fresh lava (which cools into basalt, creating a mid-ocean ridge). The elevated ridge then exerts a pushing force outward and downward on each side, away from the rift.

Both of these ideas emphasize the role of gravity. They suggest that the slightly elevated position of the mid-ocean ridges results in an outward and downward sliding movement (or push) of the plates on the opposite sides of the ridge. The plates are pushed away from the ridge, toward its opposite edge, which in many cases is a subduction boundary, where the slab-pull force will continue to drive it into the asthenosphere.

REMEMBER

While they may disagree on the particulars, scientists these days generally agree that mantle materials move by convection and that cool, dense *lithosphere* (the crust and uppermost mantle) enters the mantle at subduction zone boundaries. They have some good evidence to provide support to these ideas, as you see in the next section.

Using Convection to Explain Magma, Volcanoes, and Underwater Mountains

In Chapter 8, I note that the theory of plate tectonics is a *unifying* theory for geology. This means that the description of plate tectonics incorporates scientists' understanding of many other geologic phenomena that have been observed. In this section, for example, I explain how the subduction of crustal plates (courtesy of mantle convection) is related to volcanoes.

REMEMBER

When you're reading this section, keep in mind that the earth's mantle is solid, not liquid. A common misconception is that the mantle of the earth is made entirely of *magma*: melted, liquid rock. *It isn't!* Magma forms only under the right conditions of heat (explained in Chapter 7).

In this section, I describe three places on the earth where magma-forming conditions exist:

» In subduction zones where plates collide and one plate subducts beneath the other

» Under the crust at mantle plume hotspots (such as the Hawaiian volcanoes or Yellowstone hotspot)

» Along the boundary where two plates are moving apart (mid-ocean ridges)

Plate friction: Melting rock beneath the earth's crust

Everyone is familiar with the image of a volcano spewing (or oozing) hot lava. *Lava* is melted rock from deep in the earth's crust. Before it reaches the surface the melted rock is called *magma*; only after the volcano erupts is the melted rock called *lava*.

Melting rocks to create magma requires a certain amount of heat relatively close to the earth's surface. One place where this much heat can be created is where one plate subducts beneath another (a subduction zone). Even though the plates are at the surface of the earth and therefore cooler than materials deep in the earth, the friction or rubbing between the two plates creates extra heat — enough heat that water, carbon dioxide gas, and other elements are squeezed out of the subducted crustal rocks into the mantle and crust above them. This helps change the mantle rock to liquid magma (a process called *flux melting*; see Chapter 7 for details). The newly formed magma then moves upward through the crust. On its way to the surface, it melts some of the minerals in the rocks it touches and adds those elements to its *melt*, or liquid. Because different minerals have different melting temperatures, the magma is able to melt only certain minerals (not all of them).

REMEMBER

This incomplete melting of rocks — where only some minerals are added to the melt, while others remain solid rock — is called *partial melting*. The composition of a magma is determined by which minerals are partially melted by it as it moves upward toward the surface.

Silicate minerals (such as quartz, feldspar, and others I describe in Chapter 5) melt at lower temperatures than other minerals, so partial melting (which I describe in Chapter 7) creates magmas with high amounts of silica. When a magma containing a lot of silica cools, silicate minerals are re-formed into igneous rocks on or in the earth's crust. These rocks may, in the distant future, be subducted again (see the discussion of the rock cycle in Chapter 7), thus continually concentrating the elements that form silicate minerals and keeping them in the earth's crust.

Creating volcanic arcs and hotspots

As magma moves upward through the crust, it can cool and become an igneous rock (see Chapter 7) or it can erupt onto the surface, in which case a volcano is born! Volcanoes typically occur in two settings: 1) along the edge of a subducting plate as it partially melts, or 2) as a hotspot in the middle of a plate. I explain both settings here.

Volcanic arcs

As a plate is subducted and partially melts, a chain of volcanoes appears at the surface of the overlying plate. A volcano chain is called an *arc*. Two types of volcanic arcs are commonly recognized:

» **Island arcs:** Volcanic island arcs are created when two oceanic plates collide and one is subducted under the other. As the subducting crust is melted and the magma rises, it breaks through the crust of the overlying plate, creating a chain of volcanic islands that are roughly parallel to the edge of the subducted plate. Figure 10-2 illustrates a cross-section view of how this process occurs.

Examples of volcanic island arcs created from this process include the Aleutian Islands in the North Pacific Ocean and the Marianas Islands in the South Pacific Ocean.

» **Continental margin arcs:** When a continental plate collides with an oceanic plate, the oceanic plate is subducted beneath the continental plate (I explain why in Chapter 9). The magma rises through the overlying continental crust and creates a *continental arc:* a string of volcanoes along the continental margin, or edge. Figure 10-3 illustrates a cross-section view of how this process occurs.

Examples of continental margin arcs include the Cascade Mountains along the western coast of North America and the Andes Mountains of South America.

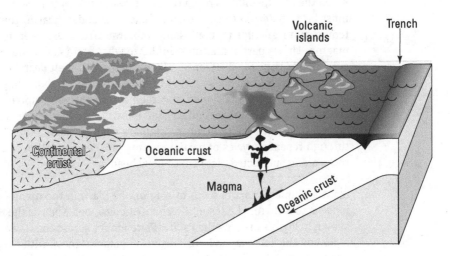

FIGURE 10-2: A cross-section view of how a volcanic island arc is created.

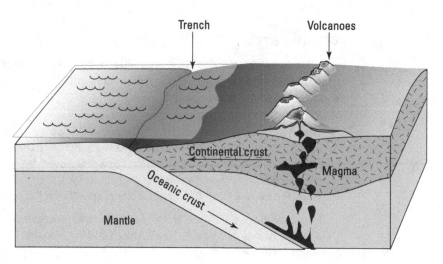

FIGURE 10-3:
A cross-section view of how a continental margin arc is created.

Volcanic hot spots

Not all volcanoes result from subduction. Some form when magma is created beneath a crustal plate due to heat released during the cooling phase of the mantle convection cycle. Remember the mantle plumes I illustrate earlier in the chapter? Sometimes these plumes of heated mantle rock move up from the earth's core toward the crust, erupting in the middle of a crustal plate. These points of eruption are called *volcanic hot spots.*

When the mantle plumes reach the crust, they cool, which means some of the heat must be transferred to the crustal rocks. The added heat in the crust raises the temperature enough to melt some minerals from the deep crustal rocks into magma. This is partial melting, similar to what occurs at subduction zones. However, there is one big difference: The minerals that melt deep in the crust are different minerals than those found (and melted) in a subducting crust. Scientists know this because the *basalt* (the rock formed when erupted magma, or *lava*, cools) at hot spots has less silica than basalt from volcanic arcs. The minerals in hot spot magmas are thought to originate deep in the mantle and have not been through repeated cycles of partial melting and concentration of elements the way rocks along subducted plate edges have been.

Mantle plume hot spots seem to stay in one place in the mantle as crustal plates move across them. The result is a chain of volcanoes, such as the Hawaiian Islands shown in Figure 10-4. As the Pacific Plate moves across the Hawaiian hot spot, the older volcanic islands (no longer erupting magma) move with the plate toward the northwest, while a new volcanic island is formed to the southeast.

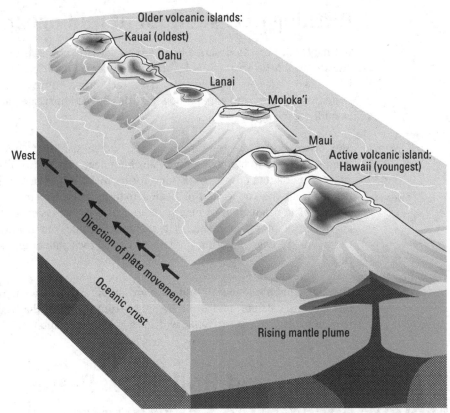

Older volcanic islands:

Kauai (oldest)

Oahu

Lanai

Moloka'i

Maui

Active volcanic island:
Hawaii (youngest)

West

Direction of plate movement

Oceanic crust

Rising mantle plume

FIGURE 10-4:
The Pacific Plate
moving across a
volcanic hot spot
created (and
continues to
create) the
Hawaiian Islands.

This process can also occur beneath continental crust plates. The geologic features of Yellowstone National Park in Wyoming (such as hot springs, which I describe in Chapter 12) are the result of a hot spot beneath the North American Plate. Geologists have recognized a sequence of progressively older *calderas* (collapsed volcanic features) stretching across southern Idaho, indicating that the Yellowstone hot spot has been active for more than 13 million years.

Remembering the ridges

Volcanoes are also formed along mid-ocean ridge boundaries where two plates are moving apart from each other, by decompression melting (see Chapter 7). These volcanoes are most often underwater and occur sporadically where magma is welling up beneath the plate boundary. One place where this type of volcano is visible is on the island nation of Iceland. Keep reading to discover more details about the processes occurring at mid-ocean ridge boundaries, like the one that splits Iceland.

Birthing new seafloor at mid-ocean ridges

In Chapter 8, I briefly describe how we can date the minerals in seafloor rocks to know how old the ocean crust is. And in Chapter 9 I describe *divergent boundaries*, where two plates are moving away from each other. In this section, I pull these two ideas together and explain how mantle convection creates new oceanic crust at mid-ocean ridges.

REMEMBER

Mid-ocean ridges are mountain ranges on the seafloor. They are created by magma welling up beneath a divergent boundary, where two plates are moving apart. The magma erupts, cools, and forms rocky ridges of basalt. The eruption results from the convection of heated mantle rocks as I discuss earlier in the chapter. The production of magma is due to decompression as the plates move apart (see Chapter 7 for details). The magma continues to rise until it erupts along the boundary edge, creating a chain of basalt rock mountains — a mid-ocean ridge.

The magma at a mid-ocean ridge has low amounts of silica and high amounts of magnesium, calcium, and iron: elements from minerals that are found in deep mantle rocks (such as *peridotite*, which I describe in Chapter 7).

Shake, Rattle, and Roll: How Plate Movements Cause Earthquakes

Whatever forces drive mantle convection, we know that convection continually drives crustal plates toward one another. It seems like the earth's crust is pretty solid and the continental plates are packed on pretty tightly, so when a plate or two start moving, they really shake things up!

Crustal rocks begin to deform anytime two plates come in contact with one another. At convergent boundaries, where plates move toward each other, the energy and tension between the two plates builds up over time. Then, SNAP! The energy is released and the ground ripples beneath us as the rocks bounce back to their original shape. Sounds pretty dramatic, and it is. This process is what we experience as an earthquake: the elastic rebound of earth's crust when built-up energy from plate movement is suddenly released.

In this section, I explain how convection-inspired plate movements result in earthquakes, how earthquakes generate waves that travel through the earth, and how we measure an earthquake's magnitude.

Responding elastically

It may seem strange to think of rocks as elastic, but that is the best way to describe how they respond to the applied pressure of plate movements. As the rocks are pressed together, they deform. Energy is stored until a *slippage* occurs, releasing the stored energy suddenly and allowing the rocks to spring back to their previous shape. This response is called *elastic rebound*.

OBSERVING EARTH'S INTERIOR BY PROXY

In Chapter 4, I discuss the interior of the earth and scientists' ideas about the solid inner core, outer liquid core, and solid (though flowing) mantle beneath the lithosphere. Did you ever wonder how scientists decided what the earth's interior looks like? Their ideas are based on observations of how seismic waves travel through the earth, so earthquakes have directly influenced what we know about the earth's layers.

Earthquake waves are recorded by instruments buried underground called *seismometers* all over the planet. When an earthquake occurs, the seismometer sends a signal to a machine in a lab (a *seismograph*) that records the earthquake wave movements on a printout called a *seismogram*. Scientists watch the seismographs as they print the seismograms to see when the P waves and S waves arrive.

Keep in mind that S waves can travel only through solid material. Scientists studying earthquakes in the early part of the twentieth century noticed that S waves didn't arrive at seismographs across the globe from their originating epicenters. The scientists determined that there was a region through which S waves wouldn't travel, and they dubbed the region the *shadow zone*. These observations have led scientists to conclude that the outer core of the earth is liquid and have allowed them to estimate at what depth this liquid layer begins.

Remember: While S waves can travel through solid material, they do not travel through the solid inner core of the earth because they are blocked (refracted) by the liquid outer core, through which they cannot travel.

P waves can travel through liquids, but they create a shadow zone of their own due to slight changes in their direction at the liquid-solid boundaries at the mantle–outer core and outer core–inner core regions of the earth. At these boundaries P waves are refracted at an angle to their original direction. They also slow down as they pass through the liquid outer core so that they arrive later than expected on the other side of the globe. Figure 4-3 in Chapter 4 shows you how S waves and P waves travel through the interior of the earth and create shadow zones.

Following the initial release of energy comes a series of *aftershocks* as the rocks continue to shift and settle into place. Similarly, as the pressure between plates builds up there may be a series of small adjustments that create *foreshocks* before the major release of energy producing an earthquake occurs.

Sending waves through the earth

If you have ever experienced an earthquake, you have felt the ground beneath you move. When the slippage occurs, the energy that is released spreads out in seismic waves. The word *seismic*, which appears frequently in earthquake discussions, comes from the Greek word for "shaking."

Two types of waves travel out from the *focus*, or origination point, of an earthquake. One type is *surface waves*, which travel across the surface of the earth just like ripples travel across the surface of a pond when you throw a pebble into it. These waves travel in all directions from the *epicenter*, which is the location on the surface directly above the earthquake's focus. Surface waves can move up and down or side to side and are responsible for most of the damage we see from an earthquake.

Another type of energy wave, a *body wave*, travels through the earth's interior. Body waves are either *primary waves* (P waves) or *secondary waves* (S waves). Here's how they differ:

» **P waves** are also called *compressional waves.* They move quickly by compressing rocks, which means the rocks contract and expand as the wave moves through them. One way to visualize this is to imagine a Slinky coil stretched out. If you compress and release one end, the energy will move along the coil by contracting and expanding until it reaches the other end. P waves can move through solids and liquids.

» **S waves** are *shear waves.* These waves move more slowly and move rocks in an up-and-down motion, very much like holding one end of a piece of rope and wiggling it up and down. Because they move more slowly, S waves arrive after P waves at a *seismograph,* or earthquake measuring station. S waves can travel only through solid material.

Measuring magnitude

When people first started "measuring" earthquakes a little more than 100 years ago, they used a scale based on eyewitness accounts of destructive damage to buildings. This is called the *Mercalli intensity scale* after its developer, Giuseppe Mercalli. Obviously, a scale like this is not very useful for measuring earthquakes in areas that do not have buildings. Since then, a Modified Mercalli Intensity Scale has been developed using modern buildings in California as its standard. However, it is useful for measuring earthquakes only in the United States and parts of Canada because many parts of the world have very different styles and standards for buildings.

Nowadays seismologists use a more accurate method of measuring earthquakes. A box called a *seismometer* is buried in the solid rock layer beneath the soil. As the waves pass through, shaking the rocks, the seismometer box also moves, and a weighted pen suspended inside (which doesn't move with the earthquake waves due to its weight and the fact that it is suspended) records the shaking movements as a series of squiggly lines. Figure 10-5 shows what a seismometer looks like.

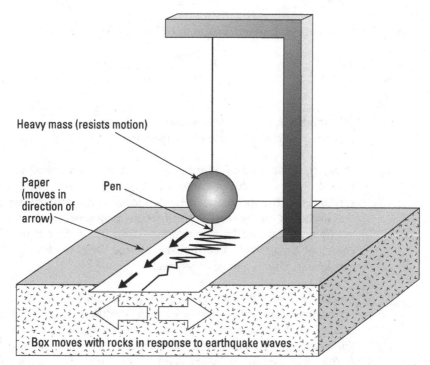

Heavy mass (resists motion)

Paper (moves in direction of arrow)

Pen

Box moves with rocks in response to earthquake waves

FIGURE 10-5:
A seismometer.

A seismometer sends a signal to a machine in a lab or monitoring station called a *seismograph*, which records the squiggly pattern of wave energy. The printout of this pattern of wave energy is called a *seismogram*. Figure 10-6 shows a seismogram of both S and P waves recorded by a seismometer.

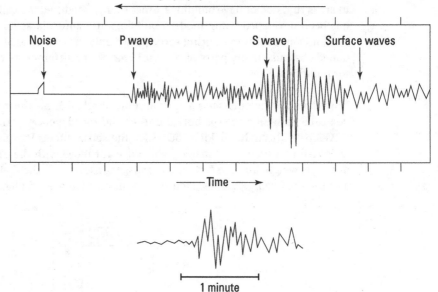

Noise   P wave   S wave   Surface waves

⟶ Time ⟶

FIGURE 10-6:
A seismogram of
S and P waves.

1 minute

Seismometers are so sensitive they can detect energy waves from earthquakes that occur on the other side of the planet!

Using the quantitative information collected from seismographs, a scientist named Charles Richter at CalTech in 1935 began to measure the magnitude of earthquakes. The Richter scale he developed is a *logarithmic scale,* which means that each unit on the scale is 10 times the magnitude of the previous value. For example, an earthquake of magnitude 4 on the Richter scale has 10 times the energy of a magnitude 3 earthquake and 100 times the energy of a magnitude 2 earthquake. That is a huge difference!

One problem with the Richter scale is that it doesn't accurately measure the really big quakes; earthquakes that measure greater than 7 on the Richter scale all tend to look similar. Therefore, scientists have developed a new measure, called the *seismic moment,* which is even more accurate. The *moment-magnitude scale* (M_w) combines measurements of the depth of the earthquake slippage and distance the rocks move with measurements of the strength of the rocks to describe the energy released by the earthquake. The calculated number quantifying the energy is then converted to a logarithmic scale comparable to the Richter scale.

4
Superficially Speaking: About Surface Processes

Learn how powerful geologic forces such as gravity and wind work with plate tectonics to create Earth's surface features.

See how water, glaciers, and waves change Earth's surface features as they move rocks and sediment from one location to the next.

IN THIS CHAPTER

» **Understanding how gravity and friction work against each other**

» **Figuring out what makes a slope stable**

» **Adding water and other factors that trigger mass wasting**

» **Moving soil and rock quickly: Falls, slides, slumps, and flows**

» **Creeping down hillsides and other slow mass wasting**

Chapter **11**

Gravity Takes Its Toll: Mass Wasting

Mass wasting is the movement of large amounts of earth materials, such as rocks, sediments, and soil, down a slope in response to gravity. (As I explain in Chapter 7, *sediments* are loose particles of weathered rock.) Also called *mass movement,* mass wasting occurs anywhere the land surface isn't completely flat. You are most likely familiar with this movement in the form of landslides and rock falls.

What causes mass wasting? It's the response of earth materials to the constant pull of gravity. *Gravity* is the force that keeps things "grounded" on earth — the same force that ensures your bread will land butter-side down when it falls off the table or promises that what goes up, must come down.

If you've ever stood at the top of a steep hill, you've experienced the pull of gravity down a slope. If you're facing downhill, to stay at the top you must use all your leg muscles and probably lean backward a little. The force of gravity encourages you to tumble forward, and it takes a good bit of energy for you to fight that pull. The steeper the slope you are standing on, the harder you have to work to resist the pull.

Gravity exerts the same pull on rocks, sediments, and soil. In this chapter, I explain what prevents earth materials from sliding down a slope. I also discuss factors that trigger mass wasting and define the various types of mass wasting.

Holding Steady or Falling Down: Friction versus Gravity

Gravity is constantly pulling materials *downslope* (toward the bottom of a slope), but a counteracting force keeps things from always, immediately sliding down. The counteracting force is *friction*, which is in a constant tug of war against gravity. When friction is winning, materials stay on the slope; when gravity is winning, materials slide downslope.

REMEMBER

Friction is the sticking force between two objects, in this case the earth materials and the surface of the slope. The amount of friction is determined by the steepness and smoothness of the slope. When the force of friction is greater than the pull of gravity, materials stay put. For example, a gentle slope or a slope with a rough surface will have strong friction with the earth materials stacked on it. Unless the slope becomes steeper or the surface is smoothed out, little to no mass wasting will occur.

However, if a slope surface is steep or smooth (or both), very little friction exists between the slope and the materials, so they can slide all the way to the bottom (where the slope flattens out).

Figure 11-1 illustrates two slopes: one (a) with a slight slope where friction is greater than gravity and the materials are stable; and another (b) where the slope is steeper and the force of gravity is greater than friction, resulting in mass wasting.

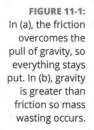

FIGURE 11-1: In (a), the friction overcomes the pull of gravity, so everything stays put. In (b), gravity is greater than friction so mass wasting occurs.

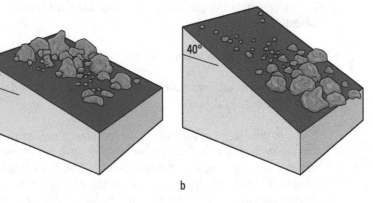

a

b

Focusing on the Materials Involved

The relationship between gravity and friction determines when materials move downslope or become unstable. Whether or not a slope is stable depends on what earth materials are involved. Soils, loose sediments, and rocks are much less stable than large sections of bedrock (solid rock). But even bedrock is susceptible to mass wasting under certain conditions. In this section, I describe how loose materials and bedrock respond differently to the pull of gravity.

Loose materials: Resting at the angle of repose

REMEMBER

In any mass wasting movement, the material moves downslope until it reaches its angle of repose, at which point it stops moving. The *angle of repose* is the angle of the slope at which sediment and rock are stable and won't move farther downslope. *Repose* means "to rest temporarily." When loose sediments are settled at this angle, for the time being they will not continue downslope; they will stay in place. However, keep in mind that the repose is temporary. Eventually, the conditions that support the current angle of repose are likely to change.

The angle of repose is different for different materials. Factors such as the sediment grain size (I explain grain size and other sediment characteristics in Chapter 7), roughness, and what the slope is made of determine the amount of friction — and, by extension, the angle of repose. Figure 11-2 illustrates how the angle of repose is different for small (fine) grains of sand, larger (coarse) grains of sand, and large, angular pebbles.

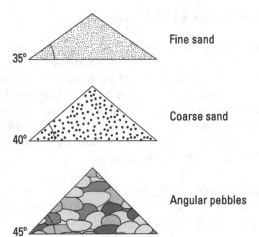

FIGURE 11-2: Sediments of different grain size have different angles of repose.

Fine sand — 35°

Coarse sand — 40°

Angular pebbles — 45°

Bedrock: Losing its stability

Beneath layers of loose sediments and soil is solid rock, or *bedrock*. Bedrock can be a sedimentary, igneous, or metamorphic rock. (See Chapter 7 for descriptions of the rock types and their formation.) Unlike loose sediments, bedrock is often very stable even on steep slopes. However, certain conditions make some bedrock susceptible to mass wasting.

Exposure to rain and wind (*weathering*) can break down pieces of sedimentary bedrock if it is not strongly cemented. Igneous and metamorphic bedrock may have *planes* (layers) of weakness that developed during formation due to how the minerals are stacked together. These areas of weakness will break apart in response to mass wasting triggers (many of which I describe in the next section), or they may develop cracks. When a bedrock has cracks, tree roots can grow into them and further break apart or weaken the rock. Similarly, water can flow into the cracks and freeze, forcing the rock to further break apart.

Triggering Mass Movements

The force of gravity is constantly pulling on earth materials, encouraging them to move downslope to a position of greater stability. When the conditions are right, large amounts of material will give in to gravity's pull. What kinds of factors can help to create the right conditions? In this section, I focus on four of them: water (which is the most common factor), a change in the slope angle, ground-shaking events (such as earthquakes and volcanoes), and the loss of vegetation.

Adding water to the mix

REMEMBER

More than anything else, water causes mass wasting events. This fact may seem counterintuitive if you have ever built a sand castle or made mud pies. In both cases, adding water makes the sand or mud particles stick together, allowing you to build a taller tower or make a stickier pie. But keep in mind that to achieve the desired effect, you have to add *just the right amount* of water. Too much water, and your sand castle slides away or your mud pie becomes mud soup. The same rule applies to sediments on a much larger scale. A little bit of water increases stickiness, but too much water leads to mass wasting.

One way water triggers mass wasting is by increasing the weight of the material involved. Adding water to soil or sediments (usually through rainfall) increases their weight, which increases the pull of gravity.

Water can also trigger mass wasting by saturating sediments. *Saturation* occurs when water fills all the space between the particles of sand or rock (called *pore space*) so that the particles can no longer stick together. When they're saturated, sediment particles no longer touch each other. Therefore, they have no friction, and they can slip right past one another. Without friction to fight the pull of gravity, sediments move downslope.

Heavy rainfall is the most common way for sediments on a stable slope to become saturated and unstable, resulting in movement downslope, or mass wasting.

Water can trigger mass wasting in multiple ways, but the water itself is not actually moving the material. *Gravity* is responsible for moving materials in mass wasting.

Changing the slope angle

Increasing the steepness of a slope increases the pull of gravity, which results in materials moving downslope until they regain the angle of repose. What could change the steepness of a slope? In many cases, water is the culprit.

Moving water in streams or rivers washes away or *erodes* sediments at the base of a slope. This type of erosion is called *undercutting,* and it creates a much steeper slope, which can trigger the movement of materials downslope.

Figure 11-3 illustrates how a stream or river washes away sediments that support the slope. This erosion by stream or river water that leads to mass wasting is called *undercutting the angle of repose.*

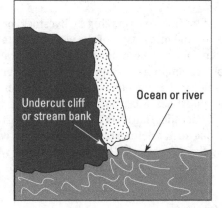
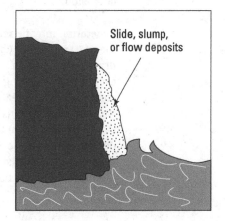

FIGURE 11-3: Stream erosion, undercutting the angle of repose, leads to mass wasting.

Is water the only factor that can change the slope angle? No — in mountainous regions, slope angles change as a result of road building. Rocky hillsides are blasted to make way for highways, and the resulting steep slopes are then prone to mass wasting as rocks succumb to the pull of gravity and fall onto the roadway below. If you ever drive through the mountains and see a sign that says "Danger: Rocks," this is exactly what the sign is warning you about. (I explain more about rockfalls in the next section.)

Shaking things up: Earthquakes

Earthquakes shake the ground and displace earth materials from their stable positions. These groundshaking events may also result in *uplift* (a change in elevation due to crustal plate movements; see Chapter 9) or fracturing of the bedrock, thus changing the slope angle, which may trigger mass wasting.

REMEMBER

If there is water present in the sediments, groundshaking events may cause the sediments to act like liquids. This process is called *liquefaction*. When liquefaction occurs, sediment particles saturated by water begin to flow like a liquid. The shaking of the sediments moves each particle away from its surrounding particles and allows water to seep in between them. When the particles aren't touching, no friction exists to compete with gravity, and thus mass wasting occurs.

Removing vegetation

The root systems of plants help to keep loose sediment and rocks in place; they stabilize the soil. In the absence of vegetation — such as when trees and plants are decimated by a forest fire — the soil and rocks are more likely to move downslope.

Vegetation may be removed by fires, overgrazing by livestock, or urban development. In regions of California and other parts of the western United States, the hot dry wildfire season is often followed by mudslides when the rains come. These events are becoming more common as certain regions experience hotter, drier conditions and greater wildfire hazards as a result of human-caused climate warming. Conditions for mudslides are also created by clear-cutting forested slopes for timber. In the last decade, Haiti has experienced devastating mudslides during hurricane season due to its deforested hillsides. (The wood is used for charcoal fuel.) Devastating flood damage in Haiti is made worse when followed by mudslides that can bury or wash away homes and people.

Moving Massive Amounts of Earth, Quickly

When the pull of gravity overcomes the force of friction, a mass wasting event is triggered. The term *landslide* is casually used to refer to any movement of earth materials by gravity. However, geologists use specific names to describe *how* rock and sediment move downslope. In this section, I describe the different types of mass wasting that occur quickly.

TIP

How do geologists define "quickly"? If earth materials are moving a few meters per second or faster, they're moving quickly.

Falls

Falls occur when rocks or sediment fall from a very steep angle or from a vertical position and move through the air before reaching the ground below. As they move, these materials do not touch the slope; they are freefalling. Figure 11-4 illustrates a fall scenario.

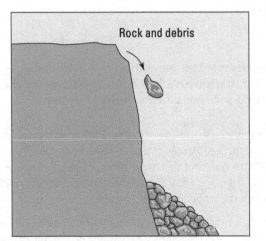

FIGURE 11-4:
Rock falls occur when materials fall through the air from a steep or vertical angled slope.

Rock and debris

Slides and slumps

The word *slide* describes the mass wasting of sediment or rock that moves downslope as one large piece. During a slide, the material stays in contact with the surface as it moves downslope.

A *slump* is similar to a slide (the materials move as one large piece), but it occurs more slowly than a slide and the materials move along a curved surface. In other words, the materials move not only down but also slightly outward, leaving a half-moon shaped scar (called a *scarp*) behind.

Figure 11-5 illustrates the difference between a slide and a slump.

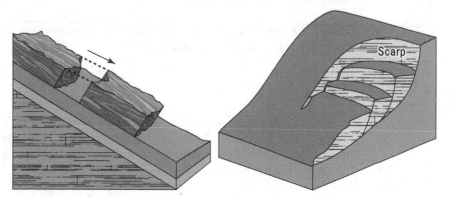

FIGURE 11-5: A slide of intact rock material and a slump leaving a scarp.

Flows

Mass wasting movements that move fluidly — like a liquid — are called *flows*. Different names are attached to various flows, which describe the amount of water involved and how the sediments interact within the flow:

>> **Earthflow:** Earthflows are relatively dry sediments with small grain sizes (usually clay or silt; see Chapter 7 for grain size descriptions) that move downslope. Of the quick-moving mass wasting types, earthflows are the slowest (at most, moving a few meters per year), but they're still faster than soil creep, which I describe in the next section.

>> **Debris flow:** Debris flows commonly occur in mountainous areas with steep slopes. They can vary in consistency depending on the mix of water and sediments; some are a thin, muddy fluid, and some are thick like concrete. Very fast debris flows down steep slopes become debris *avalanches* when the rocks and sediments fall through the air (similar to a rock fall). Debris flows composed of small sediments (such as clay and silt) are called *mudflows*.

>> **Lahars:** When volcanoes erupt, they create *lahars*, where an eruption of ash melts and mixes with glaciers or snow on a volcanic mountain and flows downhill.

A More Subtle Approach: Creep and Soil Flow (Solifluction)

Sometimes when earth materials move through mass wasting, they move very slowly. In the previous section, I describe fast mass wasting movement, which occurs at the rate of meters per second or faster. In this section, I describe slow mass wasting, measured in millimeters per year.

REMEMBER

Soil creep is so slow it can't be observed; it can be measured only over a period of time. *Soil creep* occurs as small amounts of soil are shifted downslope by the pull of gravity. Figure 11-6 illustrates that the upper layers of soil usually move more quickly than the deeper layers, so objects like trees, utility poles, or gravestones that are stuck in the soil begin to tilt downslope. The movement is so slow that trees will readjust their direction of growth and develop bent trunks in response to the creeping soil.

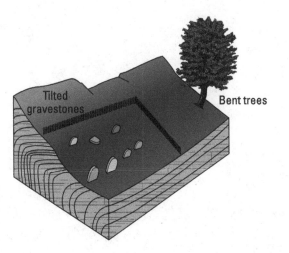

FIGURE 11-6:
As soil creep occurs, objects in the soil begin to tilt downhill.

In regions where the ground freezes in the winter, soil creep moves a little faster. *Solifluction*, or *soil flow*, occurs when the ground freezes for part of the year. In the summer, the upper layers of soil thaw while the deeper layers stay frozen (this deep frozen layer is called *permafrost*). The upper layers then become saturated with water because the water can't be absorbed by the deeper, frozen soils. The saturated layer is prone to movement due to all the water weight and reduced friction between particles.

The main types of mass wasting are listed in Table 11-1.

TABLE 11-1   **Types of Mass Wasting**

Type	Material	Speed	Water
Creep	Soils	Slow	Sometimes
Solifluction	Permafrost	Slow	Yes
Slump	Soils and sediments	Slow to fast	No
Debris flow	Sediments	Fast	Yes
Lahar	Ash	Fast	Yes
Slide	Rocks, debris	Slow to fast	No
Avalanche	Snow	Fast	No
Falls	Rocks, debris	Fast	No

Chapter **12**

Water: Above and Below Ground

Water is all around us. Most obviously, water is in the oceans — 97 percent of Earth's water is stored in them — but it's also in lakes, rivers, streams and glaciers. It flows over ground and underground, and it falls from the sky.

Flowing water is a powerful agent of change on Earth's surface. As water moves over and through Earth's crust, it picks up rocks and sediment particles, carries them far away, and leaves them somewhere else.

In this chapter, I explain the details of how flowing water on continents acts as a geologic agent. (I cover frozen water — ice and glaciers — in Chapter 13 and ocean wave movements in Chapter 15.) I begin by describing the cycling of water through oceans, air, and across continents. I explain how flowing water generates enough energy to transport rocks and other materials, as well as the features of the landscape created by flowing water. Finally, I describe how water flows underground, creates caves, and becomes heated by Earth's internal energy.

Hydrologic Cycling

The amount of water on Earth has been pretty much the same for the last billion years. Water molecules move among the earth's atmosphere, surface, and underground — and even through parts of Earth's interior — but they never leave the planet. (See Chapter 5 for a description of molecules.) The cycling of water among the atmosphere, oceans, and *lithosphere* (the rocky surface of Earth's crust; see Chapter 4) is called the *hydrologic cycle.*

The hydrologic cycle is the never-ending transformation of water into its different forms. Figure 12-1 illustrates the various parts and processes of the hydrologic cycle, which I describe in this section.

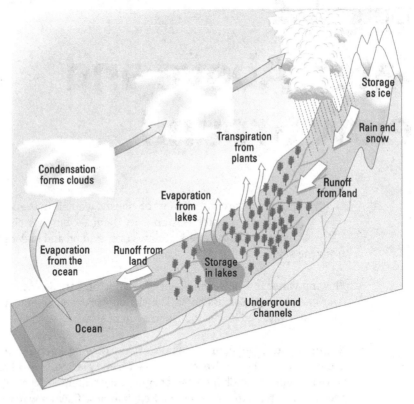

FIGURE 12-1: Earth's hydrologic cycle.

Driving the cycle with evaporation

When heated, water molecules transform from liquid to gas (water vapor or steam) in a process called *evaporation*. You observe this transformation any time you boil water.

The sun's heat fuels the evaporation of water from the ocean. In vapor form, the molecules hang out in the atmosphere together and collect additional water molecules, becoming clouds. Whenever you see clouds shape-shifting across the sky or blotting out the sun, you're actually seeing condensed water vapor that is thick enough to be visible. When the water vapor in the atmosphere is very close to Earth's surface, you call it *fog*.

Eventually the amount of water vapor in the air becomes too heavy, either because molecules are added or because the air temperature cools down. (Cold air holds less water vapor than warm air.) When the water vapor becomes too heavy, water droplets fall to the earth's surface as rain or snow (depending on the air temperature). On the surface, all this water has the same goal: to travel back to the ocean. Water can take a number of different paths to the ocean depending on where (and when) it falls.

Traveling across a continent

When clouds release rain over the continents, the water begins a long journey back to the ocean. Most water does not immediately enter a river and flow directly to the sea. Instead, it travels over ground or underground, and sometimes it becomes trapped as snow, ice, or lake water for many years before it continues to the ocean.

Here are the different paths water can take after falling from the clouds as rain (or snow):

>> **Surface runoff:** Most of the water that falls to the surface becomes surface runoff and makes its way to the nearest lake, stream, or ocean by traveling across the surface.

>> **Groundwater:** Some of the water that hits the surface is absorbed into the ground and moves through underground rocks and sediment as groundwater. This water can be accessed when you dig a well. I describe groundwater in detail in the final section of this chapter.

>> **Snowpacks or ice sheets:** In some areas, water falling as snow remains part of the *snowpack* (the snow that builds up and remains on mountaintops for many months of the year) until the next warm season, when it melts and becomes surface runoff and groundwater. Snow that falls on ice, such as at the South Pole or in Greenland, becomes part of the *ice sheet* (thick layers of ice across a large area of continent) and stays there for thousands of years.

>> **Lakes:** Water that enters a lake — either from the sky, from a stream, or as surface runoff — stays in the lake for a long time before evaporating back into the atmosphere, being carried away by another stream, or becoming groundwater.

>> **Streams:** Most of the water on the earth's surface is carried by streams. This water may begin as surface runoff, snow, rain, melting ice, or even groundwater. As it makes its way downhill, it accumulates into running bodies of water such as rivers or creeks. Geologically speaking, a *stream* is any moving body of water that carries *sediment,* or rock particles. I explain streams in detail in the next section.

It may take many years, but the water eventually makes its way back to the ocean, where once again it can be evaporated into the atmosphere by the sun's heat.

Streams: Moving Sediments toward the Ocean

TIP

People use terms such as *rivers, creeks,* and *brooks* to describe water that runs across the earth's surface, but geologically speaking these are all streams. A *stream* is any flowing water in a channel on the surface of the earth.

A stream removes or *erodes* rocks and soil, carrying sediment toward lower elevations. Eventually, a stream makes its way to the nearest ocean (or sea). On the way to the ocean, a stream drops off, or *deposits,* sediment particles. The *erosion* and *deposition* of sediments create geological features on Earth's surface. In this section, I describe the components of a stream, how streams carry sediments, and the marks they leave on the landscape when they erode and deposit sediments on their journey to the sea.

Draining the basin

A basin or *watershed* is an area of land that supplies water to a stream from rainfall and groundwater. The edges of a watershed are determined by the highest points on the landscape so that a watershed contains all water flowing downhill to a common stream that eventually empties into the ocean. Figure 12-2 illustrates how a watershed includes all the land draining toward the same direction on one side of a *drainage divide:* the highest point of elevation separating two watersheds.

Some water flows across land in a continuous thin layer, called a *sheet flow.* Some water makes its way to depressions in the landscape, which are called *channels* and become streams headed downhill. Throughout a watershed, small streams called *tributaries* flow into larger streams, such as a river or *trunk stream* that carries the water to its outlet in the nearest ocean, sea, or lake.

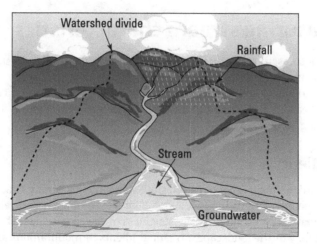

FIGURE 12-2:
A watershed.

Two types of flow

Flowing water across any surface can move in one of two ways:

>> **Laminar flow:** The streamlines of the flowing water are parallel to each other, moving in straight lines, and the water is not mixed. The surface of laminar flowing water is relatively flat and undisturbed.

>> **Turbulent flow:** The streamlines and the water are all mixed together. The surface of a turbulent flow of water is choppy and disrupted.

Figure 12-3 illustrates how the streamlines and surface in a laminar flow and turbulent flow are different.

(a) Laminar Flow

FIGURE 12-3:
(a) Laminar flow;
(b) turbulent flow.

(b) Turbulent Flow

Measuring stream characteristics

Scientists describe the movement of a stream flow using measurements of how quickly the stream moves and how much water it carries downhill. These characteristics — gradient, velocity, and discharge — are important in understanding how streams carry sediment and create the geologic features I describe in the following section.

Gradient

A stream's *gradient* is the downhill slope that it travels across. Gradient is measured as the meters of elevation lost over a kilometer of distance traveled. The gradient of a stream is determined by the steepness of the channel, which is defined by the topography of the landscape. Stream gradients in mountainous areas are likely to be steeper than stream gradients across flat, lowland regions.

Velocity

The distance water travels in a given amount of time is a stream's *velocity*. Velocity is measured as meters per second and can vary within the streamflow. Water in the stream closer to the channel edges moves at lower velocity as a result of brushing against rocks and sediment that slow it down (due to friction). Water in the center of a streamflow hits fewer obstacles and moves at higher velocity.

Discharge

A stream's *discharge* is the amount, or volume, of water that passes a given point in a set amount of time. Because discharge is a measure of volume, the width and depth of the stream, as well as the velocity (or speed), are factored into this measurement. Discharge usually increases downstream as smaller streams (tributaries) join the main channel (or trunk flow), adding more water to the stream.

Carrying a heavy load

In order for streams to carry sediment particles, they must move quickly, gathering enough energy to lift a particle into the flow. How much energy is needed depends on the size of the particle: Larger particles require faster moving water (which has higher energy) in order to be lifted and carried along. The particles being carried by a stream are called its *load*. The type of load carried by a stream is determined by the size of the sediment particle and the energy of the stream. Here, I describe the three different kinds of load.

Suspended load

Sediments that a stream carries in its flow are called the *suspended load*. These sediments are lifted away from the bottom of the stream and suspended in the moving water. When a stream appears "muddy," that means the stream has a suspended load of sediment particles. Suspended loads are most often composed of the smallest sediment particles such as clay and silt. (See Chapter 7 for details on grain sizes, clay, and silt.)

Bed load

Sediments that are too heavy to be lifted and suspended are moved along the bottom of the stream channel in the *bed load*. A stream must move quickly enough to move these particles, but they are not picked up and carried in the flow. Sediments and rocks in the bed load move along by rolling or bouncing as the water pushes them.

There are two types of bed load movement. Some sediments (usually sand-sized) move along by *saltation* or bouncing along the bottom. They are picked up by the stream just long enough to move a little bit forward before falling back to the bottom. The largest sediments (rocks or boulders) move by rolling, or *traction*, along the channel bottom.

Dissolved load

Sometimes when water moves sediments, the particles dissolve into their basic elemental ions. (I describe elements and ions in Chapter 5.) When this occurs, the particles are not visible in the streamflow. These ions compose the *dissolved load* of the stream.

Unlike suspended or bed loads, the dissolved load is not determined by the speed of flowing water. The content of the dissolved load is determined by chemical factors such as water temperature, sediment composition, and acidity of the water.

Measuring what is transported

The maximum number of sediment particles that a stream can transport is called its *capacity*. The capacity of a stream is determined by how much water is being transported: More water means more space for more sediments and, therefore, a higher capacity. Capacity is a measure of the volume of sediment that passes a set point in a certain amount of time.

A stream's capacity is limited by the volume of water (the discharge) — not by the speed of flowing water (its velocity).

REMEMBER

The *competence* of a stream is a measure of the largest particle it can carry in its flow. Because fast-moving water can carry larger particles than slow-moving water, the competence of a stream is directly related to its speed, or velocity. The competence of a single stream will vary with temperature, precipitation, and seasonal changes.

In fact, a stream's competence is its velocity squared, so even a small increase in velocity creates a large increase in competence. This fact explains why at times of flooding a stream can go from carrying sand-sized particles to carrying trees or boulders.

Eroding a Stream Channel to Base Level

As streams flow toward the sea, they erode rocks and sediment. Three common processes erode a stream channel:

>> **Abrasion:** Particles carried in the stream (usually the size of sand grains or smaller) scrape the bottom of the channel bed, scouring it.

>> **Hydraulic lifting:** The intense pressure of fast-moving water removes sediments from the stream bed.

>> **Dissolution:** The stream flows over a rock such as limestone (see Chapter 7), and the rock material dissolves into the water.

As a stream channel erodes, it deepens. But at some point, the channel cannot be any deeper; this point is called the *ultimate base level* of the stream and is the same elevation as sea level. If the channel became any deeper than its ultimate base level, the stream would need to flow uphill to continue toward the ocean, which just isn't possible.

Sometimes before a stream reaches its ultimate base level, it reaches a *temporary base level*: a point at which it cannot erode the channel any deeper but is still above its ultimate base level. A temporary base level is reached when a stream channel hits a layer of rock that is too strong to be eroded. If you have ever seen a waterfall, you have seen a stream flowing over the edge of its temporary base level. Why is it called *temporary* if the stream can't erode the channel any further? Because if the resistant rock is removed, the stream continues to try to erode the channel down to its ultimate base level.

Seeking Equilibrium after Changes in Base Level

The surface of the earth is constantly moving up and down as a result of earthquakes or continental plate collisions (which I describe in Chapter 9) or isostatic adjustment (see Chapters 10 and 13). As a result, a stream's base level is constantly being raised or lowered. The change in base level can happen suddenly, such as in response to an earthquake, or it may take hundreds or thousands of years.

When the base level of a stream is raised, the stream's flow slows down. Slower movement (lower velocity) means that the stream can no longer carry particles as large as it was carrying before. The result is *deposition* by the stream: The larger particles are left behind. (I describe depositional landforms created by streams in the next section of this chapter.) Raising the base level can happen when sea level changes, after tectonic uplift (see Chapter 9), or even when a dam is built.

The opposite occurs if a stream's base level is lowered: The stream's velocity increases as the water flows more quickly downslope. The faster movement means the stream can pick up and carry larger particles than it was carrying before. The result is increased erosion.

Raising a stream's base level leads to deposition, and lowering it leads to erosion.

If a stream were at *equilibrium*, the stream would have the exact amount of energy needed to transport all the sediment that is supplied to it. The streamflow would be neither eroding nor depositing sediment, just moving supplied sediments to the ocean. A stream in this state of equilibrium is called a *graded stream*.

The changes in velocity and competence that take place when base level is raised or lowered are a response of the stream seeking equilibrium; it wants to be a graded stream. However, the idea of a graded stream is conceptual; it doesn't often occur in real life. Streams seldom, if ever, become graded. And any stream that does become graded does not remain at equilibrium for very long because too many other factors are constantly changing, such as sediment input, water input, sea level changes, tectonic changes, and man-made obstacles.

Leaving Their Mark: How Streams Create Landforms

All this moving of sediment by streams leaves its mark on the landscape. Every time a sediment particle is picked up or set down by a stream, the surface of the earth changes just a little bit. In the previous sections, I focus on flowing water and its characteristics. In this section, I describe the sediments that are left behind, including what types of landforms streams leave on the landscape and how sediment erosion and deposition create features and shape the surface of the earth.

Draining the basin

As flowing water moves toward lower elevations and removes sediments along the way, it etches a pattern into the landscape. What the pattern looks like is determined by the characteristics (such as composition and structure) of the rocks and sediments that the stream flows over. These *drainage patterns* are described from a bird's-eye view: a view from above the land surface. Figure 12-4 illustrates the four most common drainage patterns:

» **Dendritic:** These tree-like drainage patterns develop when a stream moves across sedimentary or igneous rocks that are fairly flat. (See Chapter 7 for descriptions of rock types.)

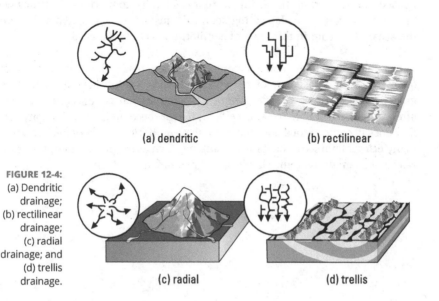

(a) dendritic   (b) rectilinear

(c) radial   (d) trellis

FIGURE 12-4: (a) Dendritic drainage; (b) rectilinear drainage; (c) radial drainage; and (d) trellis drainage.

>> **Rectilinear:** This drainage pattern looks like a blocky maze with many right-angled turns or rectangles. It's created when a stream flows across a series of fractures or faults in the bedrock. (See Chapter 9 for a description of fractures and faults.) The stream flows into the cracks already created in the rocks and erodes them further.

>> **Radial:** A radial drainage pattern occurs when a stream flows down a peak in the landscape, such as a volcanic cone (which I describe in Chapter 7). The water moves down the slopes of the peak and radiates out in all directions.

>> **Trellis:** Trellis drainage patterns occur when a stream moves through a landscape of valleys and ridges where the ridges are more resistant to erosion than the valleys. The stream flows through the valleys, parallel to the ridges, until it finds a spot that can be eroded through the ridge into the next valley.

Meandering along

Zoom in from your bird's-eye view of the drainage patterns, and you can see that a stream channel has a noticeable pattern too. Three types of stream *channel patterns* exist: braided, meandering, and straight. All three may occur within one stream's drainage at different locations depending on changes in the sediment and water volume.

Braided streams

Braided channel patterns, or *braided streams*, occur when a stream's flow is separated into multiple channels that weave back and forth, intertwining with one another. The channels are separated by mounds or *bars* of sand and gravel (that sometimes have small trees and shrubs growing on them). This pattern is illustrated in Figure 12-5. Braided streams are most common when streams have very high inputs of sediment but move slowly so that most of the sediment is deposited, creating the sand and gravel bars.

Meandering streams

A stream that flows in a single channel but winds its way across land like a snake is called a *meandering stream*. Meandering streams move back and forth across a relatively flat surface, creating loops. The continued removal and deposition of sediment in meandering streams may eventually cut off a loop from the main channel, creating an *oxbow lake*, as illustrated in Figure 12-6.

Oxbow lakes are created because of the way the stream's flow speed (velocity) changes as it twists and bends around the curves.

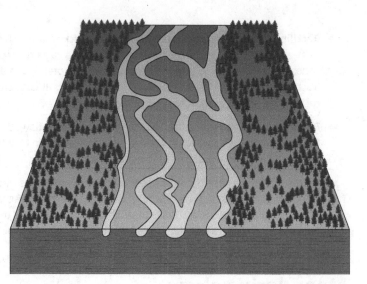

FIGURE 12-5:
A braided stream
channel.

Deposits of silt and clay

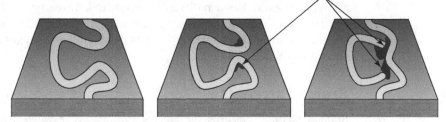

FIGURE 12-6:
A meandering
stream channel
that creates an
oxbow lake.

REMEMBER

Within any stream the velocity is greater on the outside edge of a turn, which means the outside edge picks up or erodes more sediments. Velocity is slower on the inside edge of the curve, leaving behind or depositing sediments. These deposited sediments create a *point bar.*

In the case of an oxbow lake formation, the point bar continues to grow, leading to increasingly large, bendy curves with only a narrow section of dry land between them. The next time the streamflow is greatly increased (perhaps due to rainfall upstream), a new, straighter channel will be created, cutting off the loopy edges of the old meandering channel. These old loops become oxbow lakes alongside the main river channel.

Straight stream channels

Straight channel patterns are not very common but do occur — usually where the landscape is steeply sloped and a stream flows rapidly, straight downhill. They can also appear when the streamflow is temporarily increased, causing a more rapid flow through sediments that are easily eroded.

Mount St. Helens is a volcanic mountain in southwestern Washington State that erupted explosively in 1980, creating an 8-mile radius of destruction.

A sample of porphyritic andesite, a volcanic rock with large minerals of plagioclase feldspar (white) and hornblende (black) in a matrix of smaller mineral crystals (gray); see Chapter 7.

A piece of obsidian, a volcanic rock with glassy texture and conchoidal fracture. The outside of the rock has been weathered, resulting in a brownish red color (see Chapter 7).

Vesicular basalt, a volcanic rock that cools with gases trapped inside. These gases later escape, leaving holes or vesicles in the rock (see Chapter 7).

A sample of migmatite, a metamorphic rock with light and dark bands of minerals and folding due to exposure to high temperatures and intense pressure deep within the earth's crust (see Chapter 7).

A selection of minerals (see Chapter 6): (top left) native copper; (top right) azurite; (middle left) two forms of pyrite; (middle right) calcite; (bottom left) quartz crystal; (bottom right) purple fluorite.

A sample of mica schist, a metamorphic rock composed of biotite minerals (see Chapter 7).

A close-up image of a polished piece of granite, an igneous (plutonic) rock with many similar-sized crystals of orthoclase (pink), biotite (black), albite (white), and quartz (gray); see Chapter 7.

A cavern at Carlsbad Caverns, New Mexico, with very large stalagmites (see Chapter 12).

A glacially carved U-shaped valley in Rocky Mountain National Park, Colorado (see Chapter 13).

The view to the southwest through a rift in Iceland's Þingvellir National Park. This geologic feature is a continuation of the undersea Mid-Atlantic Ridge (see Chapter 9).

The San Andreas Fault in California — a transform fault where the North American Plate and Pacific Plate slide past one another (see Chapter 9).

Fossil stromatolite (see Chapter 18) from the Whetstone Mountains, Cochise County, Arizona.

Modern stromatolites growing in Shark Bay, Australia.

Several trilobite fossils in a rock from Morocco (see Chapter 19).

Fossils of an extinct cephalopod called an ammonite, related to modern nautilus and squid (see Chapter 19).

A piece of limestone containing many fossils from the late Paleozoic (see Chapter 19) sea bottom, including a brachiopod (the large shell), bryozoa, crinoid stems, and sea urchin spines.

© ROGER WELLER

The skeleton of a Dimetrodon, a sail-backed pelycosaur common during the Permian period, about 275 million years ago (see Chapter 19).

The fossilized skeleton of confusciousornis sanctus, an ancient bird species found in China (see Chapter 20).

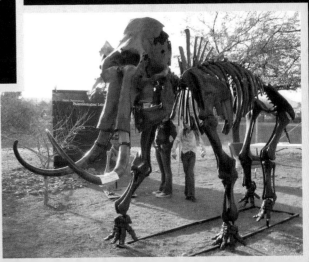

© ROGER WELLER

A Mammoth skeleton from Siberia. Mammoth and other large mammals roamed the continents during the Cenozoic (see Chapter 21), until their

Depositing sediments along the way

Any earth material that a stream leaves behind is called *alluvium*, regardless of its size or shape. Most alluvium is left somewhere on the continent as the stream moves toward the sea, with only a little bit actually being deposited into the oceans.

When a stream floods, water flows out of its channel across the stream's *floodplain*: the area on either side of the stream channel. The increased speed and volume of a flood allow larger amounts of sediment to be carried in the streamflow. But after the water spreads across the floodplain, the speed slows down and begins leaving sediments behind, creating these landforms:

>> **Levees:** Levees are created when the largest particles are deposited from the flooding stream. These particles are usually deposited right along the channel because the heaviest particles are dropped first as soon as the streamflow begins to slow down.

>> **Backswamps:** Beyond the levees, backswamps are composed of the smaller sediment particles such as clay and silt that are carried farther along as the floodwater slows down. Backswamps are usually flat areas and may have standing water after the flood waters disappear.

>> **Alluvial fans:** Occasionally, a stream flowing through a mountainous region suddenly flows out onto the relatively flat surface of a valley floor. The speed of the streamflow slows down very quickly and deposits a large amount of sediment, creating an *alluvial fan*. An alluvial fan is a sloping, fan-shaped landform that is created as the stream spreads out and slows down quickly, depositing sediments of all different sizes.

Reaching the sea

Eventually, a stream reaches the ocean, and any sediments it is still carrying are deposited. By this time only the smaller particles of sand, silt, or clay are left in the slow-moving flow of water. As these smaller particles are deposited, they create a triangular or fan-shaped landform called a *delta*.

In a delta, the larger particles are deposited first, and as the delta grows out into the sea, smaller particles are deposited in order of size (largest to smallest). A delta continues to grow as long as the stream is flowing, bringing sediments from the land to the ocean. The delta is the outlet of the stream; it represents completion of the stream's goal right from the beginning when it first started carrying sediments downslope, across the continent.

Flowing beneath Your Feet: Groundwater

While much of the water that falls as rain flows to the ocean in streams, a portion of it sinks through the soil and becomes *groundwater*. Groundwater makes up a small percentage of the earth's total water, but it supplies almost all the water that humans drink. Anytime someone digs a well, the water they pump to the surface (or draw up with a bucket) comes from the hidden supply of groundwater.

Groundwater doesn't flow as quickly as surface streams (because it has to move through sediments and rocks), but it does flow. In this section, I describe how groundwater flows underground, and I introduce the geologic features that result from the movement of groundwater, including springs, caves, and geysers.

Infiltrating tiny spaces underground

The movement of rainwater through the soil layers into sediment and rock layers below the earth's surface is called *infiltration*. The amount of water that infiltrates the ground depends on how much space exists between particles. The upper surface of the earth is like a sponge in this way; generally, enough space exists between particles to allow water to infiltrate. But that's not always the case. For example, layers of tiny clay particles are often so tightly packed together that there is no space between them for water to move through.

Three factors can increase the infiltration of sediments:

>> **Increasing the space between particles:** Creating more space between the sediment particles allows more water to move through it. For example, the growth of tree roots into the soil creates more space and increases infiltration. Similarly, space can be created by the activity of animals in the soil.

>> **Decreasing the slope of the ground:** A ground surface that is flat allows the rainfall to sink in before running off. If the ground surface is steeply sloped, rainfall runs off as surface or streamflow before it has a chance to infiltrate the surface and become groundwater.

>> **Lowering the intensity of rainfall:** Rainfall resupplies groundwater, but there can be too much of a good thing. Too much rain that falls too quickly may fill only the upper layers of sediment, leaving additional water to run off in streams. Think about when you water your lawn or garden: The plant roots deep in the soil get much more water if you use a gentle sprinkler for many hours than if you turn the hose on full blast for 10 minutes, creating large surface puddles but little infiltration.

Measuring porosity and permeability

Two measurements describe how much water a rock or sediment layer will hold:

>> **Porosity:** A measure of the amount of space between particles, porosity is expressed as the percentage of the total volume.

>> **Permeability:** Describes how easily water can move through the rock or sediment. It is determined by both the porosity and how well the spaces are connected (to allow flow from one space to another).

REMEMBER

A sediment or rock can be porous (having space between its particles) but not permeable if the spaces are not connected in a way that allows flow of water from one space to another.

When groundwater flows, it is responding to gravity, just like surface flow in streams. Groundwater flow is also affected by pressure from the earth materials piled above it. The combination of gravity and pressure is called the groundwater's *potential.* When groundwater flows, it flows from areas of high potential (a high pull of gravity and high amounts of pressure) toward areas of low potential. Because potential is calculated using the force of gravity, areas of low potential are commonly downhill from areas of higher potential.

TECHNICAL
STUFF

The difference in groundwater potential across a measure of distance (like a mile or kilometer) is called the *hydraulic gradient,* which indicates the direction of groundwater flow.

Setting the water table

Scientists describe two zones of rock and sediment in the groundwater system. Water moves first through a layer of sediments or soil called the *zone of aeration.* In this zone, most of the space between sediment particles is filled with air, and water can move through it. At some point, the water reaches a place where the space between sediment particles is filled with water instead of air; this is called the *zone of saturation.* The zone of saturation is what you aim for when you dig a well. Figure 12-7 is a sketch of the different layers water moves through as it becomes groundwater beneath the surface.

REMEMBER

The boundary between the zone of aeration and the zone of saturation is called the *water table.*

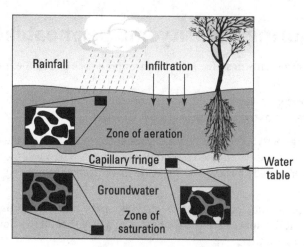

FIGURE 12-7:
Beneath the surface as water infiltrates sediment and rock layers.

At the water table, an interesting interaction occurs. While the water is being pulled downward by gravity, some water is also moving upward into the lowest part of the zone of aeration — against the pull of gravity. This upward movement is called *capillary action*. You may also call it *wicking action* because it mirrors the way a tissue or paper towel, if dipped in water, draws the water upward (or the way your pant cuffs get wet from walking on wet streets). This capillary action occurs just above the water table in an area called the *capillary fringe.*

TECHNICAL STUFF

Capillary action is the result of the attraction of water molecules to other molecules or surfaces (due to the slight polarity of a water molecule, which I briefly describe in Chapter 5). This attraction overcomes the pull of gravity, allowing the water to move upward.

Springing from rocks

Groundwater flows, for the most part, below ground. But at certain places, the water table intersects the earth's surface due to the folding of rock layers or erosion. Groundwater can appear on the surface in the form of lakes, swamps, and springs.

Springs are locations where groundwater flows out onto the surface. You often find springs on slopes where the change in elevation of the ground surface intersects the elevation of the water table, allowing groundwater to flow to the surface out of the hillside, as illustrated in Figure 12-8.

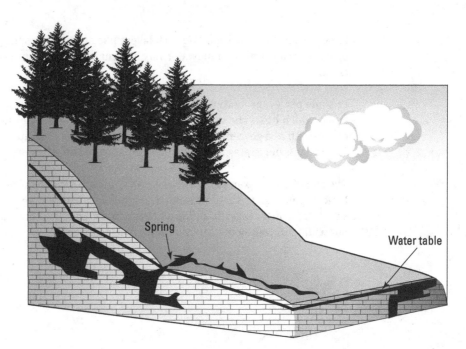

FIGURE 12-8:
Springs often occur on hillsides, where groundwater flows out onto the surface.

Spring

Water table

Confining an aquifer

An *aquifer* is any rock or sediment where fresh water is stored. The saturated zone below the water table that you tap into with your well is one example of an aquifer. It is an *unconfined* aquifer, which means that the rocks and sediment above the stored water are permeable (water can flow through them).

Confined aquifers, or *artesian* aquifers, are layers of rock or sediment filled with water that have an *impermeable* layer above them. This means that water does not easily infiltrate down to a confined aquifer directly from the earth's surface; it can't pass through impermeable layers. Instead, water enters a confined aquifer by flowing horizontally through other permeable rocks or sediment.

An interesting feature of artesian aquifers is that due to the pressure of the water being trapped between two impermeable layers, the water may be forced upward against the pull of gravity. The next time you buy bottled water from an "artesian spring," you'll understand that the water flowed up to the surface from a confined aquifer beneath the earth's surface!

Heating up underground: Geysers

A *geyser* is like an artesian spring: Groundwater flows up to the surface, against the pull of gravity. But a geyser is not forced upward due to pressure and the impermeability of surrounding rock layers like an artesian spring. A geyser flows upward because the groundwater has become very hot from heat in the earth's crust.

In some places, groundwater flows close to magma moving upward in the earth's crust, which heats the water. After it's heated, the water rises toward the surface of the earth as a geyser. (I explain why heated materials rise upward in Chapter 10.) Figure 12-9 illustrates how groundwater is heated by exposure to *magma*.

The most famous example of a geyser is Old Faithful located in Yellowstone National Park, Wyoming. It shoots heated water up into the air on a fairly predictable schedule, which makes it an ideal attraction for people curious about geysers but not willing to wait hours (or even days) in the hope of seeing an eruption.

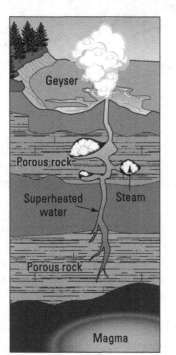

FIGURE 12-9:
Groundwater heated by magma rises to the surface as a geyser.

That sinking feeling: Karst, caves, and sinkholes

When rocks and sediment made of certain minerals (see Chapter 6) contact water, the minerals may dissolve and be carried away in the water flow, leaving holes in the rock. Such rocks are said to be *soluble*. When groundwater flows through soluble rock (usually carbonate rocks like limestone), it creates small spaces where the soluble minerals used to be. Continuous groundwater flow enlarges the spaces over time. An area of the earth's crust that has been eroded by groundwater in this way is called *karst*.

Features of karst are visible above and below the ground. The most common are *caves*. Groundwater creates caves by flowing through the soluble rock and dissolving holes into it. At some point after this occurs, the water table level changes. The large holes in the rock remain, but they're no longer filled with water. Very large caves, or systems of connected caves, are called *caverns*. The interior of a large cavern is featured in this book's color photo section.

Drip, drip, dripping: The formation of dripstones

Dripstones, found inside caves and caverns, form as water drips from the cave ceiling. They look like icicles of rock, growing longer with each drop. They grow because the dripping water contains dissolved minerals that it leaves behind.

Two types of dripstones exist:

>> **Stalactites:** These icicle-like dripstones grow from the ceiling of a cave.

>> **Stalagmites:** Stalagmites grow upward from the floor of a cave, directly beneath stalactites. The same dripping water that creates stalactites falls down to the cave floor, depositing minerals that form a stalagmite.

Eventually, the stalactites and stalagmites grow together, meeting in the middle and forming a *column*.

TIP

One way to recall the difference between a stalactite and stalagmite is to remember that stalactites hang "tite" and you "mite" trip over a stalagmite.

Sinkholes and disappearing streams

In karst landscapes, the abundance of caves below the surface may cause a *sinkhole*. A sinkhole is created when the rock below the soil is dissolved by groundwater and the soil sinks to fill the space, creating a depression. Sinkholes may also result from the roof of a cave collapsing, leaving a depression in the ground surface.

TIP

When you see a sinkhole, you can be certain the underlying rocks are soluble. In the United States, sinkholes are common in Florida, Georgia, North Carolina, and South Carolina. They can also be found in many different regions of the world where the rocks below the surface are soluble, such as limestone.

The presence of sinkholes causes *disappearing streams.* Disappearing streams flow across the surface for only a short distance before they encounter a sinkhole. The sinkhole has changed the slope of the ground surface by sinking downward. The stream flows down into the sinkhole and "disappears" below ground, continuing to flow, but going through caves and underground spaces left in the soluble rock by previous groundwater.

Chapter **13**

Flowing Slowly toward the Sea: Glaciers

Relative to the size of the oceans, very little of the earth's water is ice. Currently only 10 percent of the planet's water is stored in glaciers, ice caps, and ice sheets. However, at times in the past, the amount of ice covering the continents and oceans was much greater. Scientists know that ice covered portions of some of the continents because of evidence — left behind like fingerprints — that could be created only by flowing ice.

Wait, *flowing* ice? Yes, ice flows — similar to, yet also very different from, the way a liquid flows. Massive amounts of ice such as glaciers, while solid, move by what geologists call *solid* or *plastic flow* — terms I define in this chapter.

Here, I describe how glaciers form, explain the details of how glaciers flow, and illustrate the different features of the landscape created by glaciers. I also explain what causes long periods of *glaciation* (when ice covers portions of the continents and sometimes oceans) and how the presence (or absence) of glaciers affects sea level.

Identifying Three Types of Glaciers

Glaciers are massive amounts of ice — somewhat like large, frozen streams. In order for glaciers to form, they need cold temperatures and high amounts of snowfall. These conditions occur most often at *high latitudes* (areas closest to the North or South Pole) or closer to the equator at *high altitudes,* such as in mountain ranges.

Scientists categorize the ice on Earth into the following types:

>> **Valley glaciers:** *Valley* or *alpine glaciers* are the most common glaciers and the smallest. Thousands of valley glaciers exist in mountain ranges all over the world at high and low latitudes. Valley glaciers flow in a path similar to streams — through mountain valleys toward lower elevations and (eventually) the sea.

>> **Ice sheets:** Only two large, continent-covering ice sheets exist today: one covering Greenland, and one covering Antarctica. *Ice sheets* are not contained by valleys and flow out in all directions from a central point. When ice sheets flow out over the sea, they form *ice shelves.*

>> **Ice caps:** *Ice caps* cover areas of a continent, such as portions of Iceland, on a much smaller scale than ice sheets. Like ice sheets, ice caps flow out in all directions (The Arctic ice cap is technically not a cap because it covers only ocean. Instead, it is a mass of *sea ice.*)

Sometimes an ice sheet or ice cap has valley glaciers flowing from its main body into surrounding valleys. These are called *outlet glaciers* because they are outlets for the main ice sheet body. When many alpine or valley glaciers flow together and form a large, broad glacier that spreads out in the lowlands at the base of a mountain, it is called a *piedmont glacier.*

Understanding Ice as a Geologic Force

Glaciers combine the flowing movement of a liquid with the strength of a solid, and the result is an incredibly powerful force for moving rocks and sediment on the earth's surface. In this section, I describe how scientists track the growth and melting of glaciers, and I explain the basic mechanics of how glacial ice flows.

Transforming snow into ice

When snow falls on a glacier, it becomes packed down by the pressure of added snow on top of it. Over time this pressure and compaction force the snowflakes closely together, reducing air space between them and creating a tightly packed, dense snow called *firn*. Firn resists completely melting in the summer and refreezes each winter. The annual melting and refreezing continue to reduce the space between snow particles until no air is left, only solid ice — just like the ice cube in your glass.

Balancing the glacial budget

One way scientists try to understand glaciers is through a model (I explain the use of models in science in Chapter 2) called the *glacial budget*. The glacial budget is a way to think about how a glacier is growing or shrinking. Just like calculating your household budget, the glacial budget looks at the additions and subtractions throughout the year, but in this case ice is added or subtracted, not dollars.

Each year a glacier goes through a cycle of addition and subtraction in response to the changing weather. In the winter months, when temperatures are low and snowfall is high, glacier ice is added, or *accumulates*. As the temperatures warm and snowfall decreases in the spring and summer months, glacier ice is subtracted, or *ablates* by evaporation, melting, or breaking off at the downslope end. (When ice breaks off from glaciers into the ocean, the process is called the *calving of icebergs*.)

Figure 13-1 illustrates zones of ice accumulation and ablation in the glacial budget model. The *zone of accumulation* is at high elevation, where more snow falls each winter than is melted in the summer. The *zone of ablation* is at a lower elevation, where all the snow that falls each year melts (along with some of the ice from previous years). The *line of equilibrium* or *snowline* separates these two parts of the glacier.

When a glacier's annual accumulation of ice is more than its ablation, or *wastage*, for a few consecutive years, it has a positive budget and is increasing in size or *advancing*. As a glacier advances, the *terminus* — the edge of the glacier at lower elevation — moves downslope.

In the reverse situation — when the glacier experiences more wastage, or ablation, than accumulation — the glacier has a negative budget. The terminus of a glacier with a negative budget will *retreat*, or move upslope, as it melts and shrinks. After many years of warm temperatures and minimal snowfall, a glacier may completely disappear.

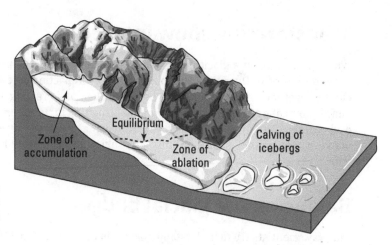

REMEMBER

The ice in a glacier always flows downhill. When you read that a glacier has retreated, it does not mean that it flowed upslope. Rather, it means that the lower end, or terminus, of the glacier is now located further upslope than it was previously. This is due to loss of ice through melting and evaporation at the terminus, not to the glacier flowing backward up the slope.

Glacial retreat and disappearance are signals scientists have observed to document modern climate warming. Glaciers in many parts of the northern hemisphere are retreating, and some have disappeared altogether.

Flowing solidly down the mountain

Regardless of whether a glacier is advancing or retreating, the ice of the glacier is always flowing downhill — from the zone of accumulation toward the zone of ablation.

It may seem strange to think of a solid moving or flowing. The flow of a solid is different from the flow of a liquid. In a solid, the atoms (see Chapter 5) are closely packed together and don't have any room to move freely like they do in a liquid or gas. Glacier ice moves in two ways:

REMEMBER

>> **Plastic flow** or **internal deformation:** In order for a solid to experience *plastic flow*, the interlocked molecules (in this case, ice crystals) shift together under intense pressure. In a glacier, the pressure comes from the weight of the ice piled above; it forces the ice crystals to slip past one another.

The rate of plastic flow in a glacier increases as the temperature of the ice approaches the melting point — the warmer the ice, the more quickly it flows.

>> **Basal sliding:** Glaciers also experience some melting of ice along the bottom of the glacier where it contacts the ground surface. Friction of the ice against the bedrock produces a small amount of heat and results in a thin layer of meltwater. The water from the melting helps the glacier slip-slide along the ground, a movement called *basal sliding*.

As the glacier moves, the upper surface of the glacier is not subjected to the great amounts of pressure that cause the internal ice to flow. The surface ice is brittle and responds to the movement of the glacier by cracking and breaking — a response that you more typically expect from a solid. As the internal ice flows, cracks or *crevasses* are created in the surface of the glacier.

REMEMBER

The rate of glacier flow is slow compared to the flow of water: Large ice sheets may flow only a few feet or meters per year, while alpine glaciers in warmer regions may flow a couple hundred meters in a year.

Occasionally, a glacier experiences a *surge* when its flow increases greatly for a short period of time, such as 100 feet per day for several months. Scientists think a surge occurs when the water underneath the glacier builds up and leads to a sudden increase in basal sliding.

Eroding at a Snail's Pace: Landforms Created by Glacial Erosion

In Chapter 12, I explain how water carries rocks and sediments in streams. The power of a stream to carry earth materials is called its *competence*. Because it is solid, ice has a very high competence for carrying rocks and sediment. This fact makes glaciers a major mover of earth materials — able to carry large amounts of sediment and very large boulders over great distances.

Today, glaciers are limited to a relatively small portion of the earth's surface, but in the past when glaciers were much more extensive they shaped the landscape by moving rocks and sediment. In this section, I describe the different ways glacial ice removes (erodes) earth materials and the features of the landscape created by glacial erosion.

Plucking and abrading along the way

Glaciers remove rocks and sediment in two ways:

>> **Plucking:** Plucking (or *quarrying*) occurs when ice moves across a surface of bedrock and lifts large blocks of rock up into the ice flow. The ice plucks the rocks by catching the edge of a previously created fracture (likely created by water seeping into cracks and freezing) and applies enough force to break off a chunk of rock. The rock is then carried in the ice flow until it is dropped or reaches the *terminus* (end) of the glacier and melts out.

>> **Abrading:** When a glacier carries rocks and sediments, some of these materials scrape against the surface of bedrock that it flows over. This scraping is called *abrasion*. The process of abrasion is similar to running sandpaper over a surface: It leaves long, straight scratches in the surface (called *glacial striations*) and can produce a polished appearance on the bedrock. The result of abrasion depends on the size of rock or sediment particles carried in the ice: Larger particles and rocks create striations and grooves, while smaller particles create a polished appearance.

TIP

Geologists can look at the striations left behind by glacial movement and determine which direction the ice was flowing — even when the glacier melted and disappeared long ago. In Chapter 8, I tell you how this type of evidence was used to develop the continental drift hypothesis.

As the rocks in the ice grind against and scratch the surface of the bedrock, a very fine, powder-like sediment is produced called *rock flour*. The milky appearance of streams fed by melting glacial ice is a result of all the rock flour being carried in the water.

Creating their own valleys

Unlike streams, which create valleys by removing sediments and rocks with their flowing water (see Chapter 12), glaciers follow existing stream valleys and carve them into much larger, wider shapes. Streams create valleys that are V-shaped with the narrow stream channel as the base, or point, of the V. Glacier ice changes these V-shaped valleys into U-shaped valleys.

The U-shaped valley of a glacier is called a *glacial trough*. Glacial troughs are wider, deeper, and straighter than stream channel valleys because the strength of the ice shears off so much rock and sediment as it pushes its way downslope.

Glacial troughs that reach the sea at the edge of a continent are filled in as small, steep-sided bays, or *fjords*, when the glacial ice no longer fills the valley.

The larger the amount of ice in a glacier, the stronger its force of erosion will be. As small tributary glaciers feed into a main or *trunk glacier*, the trunk glacier, being larger, will cut its valley more deeply than the surrounding tributary glacial valleys. When the glaciers retreat, the tributary glaciers leave *hanging valleys* along the path of the trunk valley. The position of hanging valleys is illustrated in Figure 13-2, along with other erosional features created by glaciers that I describe in the next section.

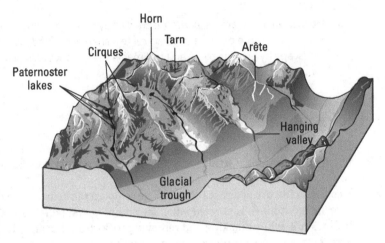

FIGURE 13-2: Landscape features of alpine glacial erosion.

Speaking French: Cirques, arêtes, et roche moutonnées

Scientists began studying glacial landscapes in the Alps of Europe. That's why many of the words describing features of glacial erosion have French origins. So when you talk about glacial landscapes, you'll find yourself speaking French!

Alpine glacial erosion

Features of alpine glaciation are what remain on any mountainous landscape after glacier ice has retreated or melted. Many of the jagged mountain landscapes you see today are the result of glaciation that occurred thousands of years ago. The features I describe here are labeled in Figure 13-2.

>> **Paternoster lakes:** These lakes are created in the glacial trough as ice plucks pieces of bedrock from the valley floor, leaving a string of holes that are later filled by water and become lakes.

>> **Cirques:** *Cirques* are the circular (or half-circle shaped) depressions near a mountaintop where alpine glaciers begin. Cirques are rounded or U-shaped, scoured features that often fill with water and become cirque lakes (called *tarns*) after the glacier disappears.

>> **Arêtes:** When multiple glaciers exist on a mountain, two cirques (or two *moraines*, which are described in the next section) may create a sharp, steep ridge of erosion-resistant rock between them. This linear ridge is called an *arête.*

>> **Horns:** Similar to arêtes, *horns* are created by multiple cirques on a mountain. In this case, three or more cirques create a sharply pointed peak in the landscape by eroding rocks and sediments from many sides of a mountaintop.

Ice sheet glacial erosion

Ice sheets are much larger than alpine or valley glaciers and, therefore, leave much larger erosional features on the landscape when they disappear.

A common feature where ice sheets have sculpted the bedrock is a *roche moutonnée.* A roche moutonnée is created when ice smooths and polishes the uphill side of a bedrock hill, while plucking and removing rocks from the downhill side as it flows across. This process is illustrated in Figure 13-3.

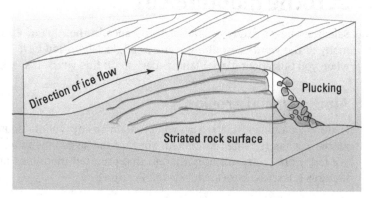

FIGURE 13-3: How ice sheet flow creates a roche moutonnée.

Leaving It All Behind: Glacial Deposits

In the previous section, I describe features created by the removal of rocks and sediment by glacier ice. All that earth material carried by glaciers has to end up somewhere. In this section, I describe landscape features created by the *deposition*, or leaving behind, of earth materials by glaciers.

Because glacier ice can carry large amounts of sediment and large sized rocks, when these materials are left behind they create easily recognizable features. Such features tell us that ice once covered the land in places you would not imagine ice existing today, such as New York City or London, England.

Glacial drift describes any sediment left behind by melting glacial ice. Before scientists understood how ice sheets covered the land, they proposed the hypothesis that icebergs had drifted these materials into their present location and melted, leaving them behind. They were only partly correct. Glacial drift is the result of movement and deposition by ice but not by floating icebergs; these sediments were deposited by ice sheets and glaciers.

Depositing the till

Glacial till is any rock or sediment that has been carried and left directly by glacial ice. Because ice can carry rocks and sediment of various sizes without separating them by size, till deposits are often a mix of different sized rocks and sediment.

These mixed deposits are called *poorly sorted* and are different from the well-sorted deposits (those of similar-sized rocks or sediments) typical of wind or water deposition (see Chapters 12 and 14).

Moraines

Glacial till often piles up alongside the edges of a glacier or at the end of one. These piles of poorly sorted sediments are called *moraines. Terminal moraines* or *end moraines* occur at the end of a glacier as it recedes, melting and leaving the till behind. *Lateral moraines* occur along the edges of a glacier. *Medial moraines* occur in between two glaciers (where two lateral moraines come together). Each of these moraine types is illustrated in Figure 13-4.

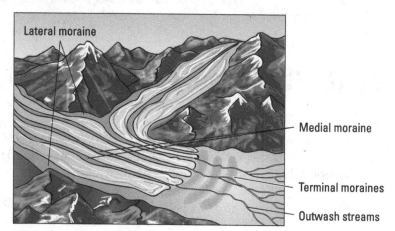

FIGURE 13-4:
Different types of glacial moraines.

Drumlins

When a large glacier or ice sheet advances over preexisting moraines (from the earlier advance and retreat of glaciers), the original till deposit will be reshaped by erosion into rounded, elongated hills called *drumlins*. The narrow, pointed end of the drumlin (shown in upcoming Figure 13-5) points in the direction of the ice flow that shaped it. Drumlins created by ice sheets often occur in groups, or *drumlin fields*, with hundreds of drumlins scattered across the landscape.

Plains, trains, eskers, and kames

Stratified glacial deposits occur when sediment is removed from the glacial ice by meltwater in the form of streams. The streams originate in or below the glacier as it begins to melt, usually near the end of the glacier. The meltwater flows farther downhill, beyond the terminus of the ice.

Deposits left behind by meltwater are well-sorted because they are transported by water. They are deposited by size (the largest settling first) in layers or strata. Landscape features made of stratified deposits are composed of particles small enough to be carried by water, such as sand or clay (I describe grain sizes in Chapter 7). Figure 13-5 illustrates the landscape features I describe in this section.

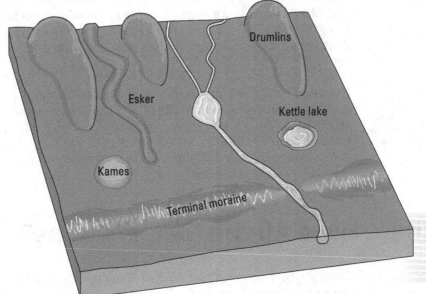

FIGURE 13-5:
Features of glacial
deposition.

Outwash plains

Just beyond the terminus of a glacier, there is often a braided stream (see Chapter 12) created by water melting from the glacier. In a mountain valley, the deposits left by this stream are called a *valley train*. When these deposits are left beyond the edge of a melting ice sheet, they create an *outwash plain*.

Sometimes, as the ice retreats, it leaves large blocks of ice sitting in the sediments of the outwash plain. When these blocks melt they leave a water-filled depression in the landscape called a *kettle lake*.

Eskers and kames

Eskers and *kames* are deposits of sand and gravel left by meltwater flowing in, under, or over the ice of a melting glacier. Eskers are long, snake-like ridges of sand and gravel. Kames are steep-sided hills created by the deposition of sand and gravel in depressions on top of melting ice. When the glacier eventually disappears, a pile of sand and gravel is left.

Behaving erratically: Large boulders in odd places

An ice sheet or glacier carries large boulders long distances and then melts, leaving these boulders in places where you least expect them. The stranded rocks are called *glacial erratics*. Scientists recognize glacial erratics because they're large chunks of bedrock that are different from the bedrock they're deposited on. For example, they may find a large boulder of metamorphic rock in a landscape with igneous bedrock. (See Chapter 7 for descriptions of the rock types.) In areas where these rocks dot the otherwise flat, rock-free landscape, they are often piled up to create fences between farm fields.

TIP

If you ever encounter a large boulder than looks as if it must have fallen out of the sky (for example, there are no rocky cliffs nearby), you've probably found a glacial erratic. In the Puget Sound region surrounding Seattle, Washington, many of the glacial erratics deposited 12,000–14,000 years ago are far too large to be moved and have been incorporated into small neighborhood parks rather than removed to make way for development.

Where Have All the Glaciers Gone?

Scientists have observed that modern warming due to human-caused climate change is melting today's glaciers and ice caps. Evidence across the landscape tells a story of ice long ago covering continents and reaching much closer to the equator than it does today. Is modern global warming responsible for these changes? And what can scientists learn from studying the glaciated landscape of the past?

In this section, I briefly describe what is known about the history of glaciers and ice sheets on Earth, how the continents respond to disappearing ice, and what the future may hold as the ice continues to melt.

Filling the erosional gaps

Scientists called *glacial geomorphologists* study the landscape shaped by glaciers and ice sheets to understand how much ice covered the continents and when it retreated or melted. Periods in the past when ice covered much of the continents are called *ice ages*. How many ice ages the earth has experienced and the geographic extent of the ice sheets are just two of the questions scientists try to answer.

Clues left in the deposits help answer these questions like pieces of a puzzle. How-ever, solving this puzzle is challenging because much of the landscape created by glacial erosion and deposition has since been eroded further by other geologic agents. This subsequent erosion creates *erosional gaps* in the record of ice ages: periods of glacial history from which the evidence has been carried away, moved, or reshaped from its original form of deposition.

Fortunately, sediments from the ocean floor also contain evidence for ice ages and the periods of warming and melting in between (called *interglacials*). Scientists combine information from the seafloor sediments with glacial features on land to estimate the time and extent of glacial ice ages.

Cycling through ice ages

For most of the earth's long history (I explain the depth of geologic time in Chapter 16), the planet has been too warm for large amounts of ice to cover the continents. In the last 3 million years (relatively recent in geologic time), periods of extreme cold have come and gone, causing ice sheets and glaciers to advance and retreat repeatedly — a cycle of ice ages.

Scientists have determined two primary causes of ice age cycling: changes in the position of the continents on the earth's surface and changes in the earth's posi-tion relative to the sun.

Moving to colder regions

When the continental plates move around the earth's surface, sometimes they move closer to the North or South Pole. (I explain the movement of continental plates in Chapter 9.) As they approach a pole, these landscapes experience wetter winters and more extreme cold— conditions that lead to the creation of glaciers and ice sheets.

The presence of ice on the land and ocean surface at the poles becomes a mirror that reflects the sun's heat instead of absorbing it. This reflection, or *albedo*, leads the earth's entire climate to become colder and intensifies the cold of winter, cre-ating larger, more extensive sheets of ice.

Orbiting, spinning, and tilting around the sun

The other cause of ice age cycles is the position of the earth relative to the sun. As a planet orbiting the sun, the earth experiences three different cycles that each occur over thousands of years. These cycles are described as *eccentricity, obliquity,* and *precession.* Together they are named *Milankovitch cycles* after the astronomer

Milutin Milanković, who first calculated them about a hundred years ago. They are illustrated in Figure 13-6, and I describe them here:

>> **Eccentricity:** As the earth orbits the sun, it does not always follow a perfectly circular route. Sometimes its route is more oval shaped, meaning that at the farthest reaches of its orbital loop it is farthest from the sun. This change occurs over a period of 100,000 years.

>> **Obliquity:** As the earth rotates on its axis and moves around the sun, it is also slightly tilted. The degree of tilt shifts about 2 degrees over a period of 40,000 years. When the earth is tilted the most, the northern hemisphere of the planet is tilted farthest away from the sun.

>> **Precession:** Along with tilting, the earth also wobbles on its axis. The axis (at the North Pole) points in different directions in the sky over a period of about 26,000 years.

Each of these three cycles occurs independently and at the same time over a period of thousands of years. When the extreme of each cycle coincides — eccentricity takes the earth far from the sun, the tilt is at its greatest, and the precession points the axis away from the sun — the northern hemisphere experiences very cold conditions, and ice builds up. Likewise, when the earth is closest to the sun, only slightly tilted, and the axis is pointed toward the sun, the northern hemisphere is very warm and has no ice present.

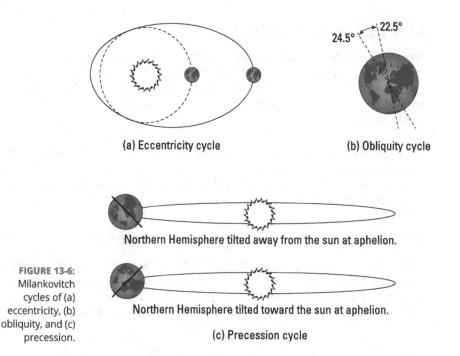

(a) Eccentricity cycle

24.5° 22.5°

(b) Obliquity cycle

Northern Hemisphere tilted away from the sun at aphelion.

Northern Hemisphere tilted toward the sun at aphelion.

(c) Precession cycle

FIGURE 13-6: Milankovitch cycles of (a) eccentricity, (b) obliquity, and (c) precession.

Milanković originally composed his hypothesis based on mathematical calculations and astronomical understanding. Since then, scientists have gathered other evidence and confirmed cycles of colder ocean temperatures occurring in sync with Milanković's predictions.

Rebounding isostatically

Each time an ice sheet advances across a continent, all that added weight of the ice causes the continental crust to sink into the mantle below just a little bit deeper. (See Chapter 9 for a description of how the crust and mantle interact.) The result of all that ice pushing down on parts of the continent changes the relative sea level along the coasts.

TIP

Scientists use the word *relative* because no unchanging point exists against which to compare changes in sea level or continent level. Any change in sea level has to be described relative to where the continent sits.

So much water is contained in the ice that you may expect the sea level to be lower, and indeed there is less liquid water in the ocean when so much of it is frozen. But at the same time, the continent is sinking downward under the weight of the ice, so the change in sea level relative to the position of the continent along some coastlines is not as dramatic as you may expect.

When the ice sheet melts, the sea level rises relative to the continent — at first. In the years following the disappearance of the ice sheet, the continent, which had been pushed down by the weight of the ice, begins to slowly bounce back up (because the weight is removed). This shift is happening right now in response to the end of the last ice age 14,000 years ago. Continents in the northern hemisphere (including North America) are slowly adjusting upward by a few millimeters a year, "floating" in the mantle — much like an ice cube that's dropped in a glass of water floats up to the top. This movement in response to the removal of the weight of the ice sheet is called *isostatic rebound.*

When you drop an ice cube into a glass of water, it doesn't just sink and then bob up to a floating position. It takes a few seconds of bobbing motion, and perhaps some tilting and shifting, before it is balanced and floating just right. As the continents experience isostatic rebound, they go through a similar period of adjustment, but over a much longer time. For this reason, some regions may appear to be moving upward while others are moving slightly downward or not moving at all.

The reason continents rebound when the ice sheet retreats is related to the density of the continental crust and the earth's mantle. (I explain density in Chapters 9 and 10). The rock that the continents are made of is less dense than the mantle rock below it, so the crust floats in the mantle like an ice cube floats in your water glass. If you add weight to the ice cube (for example, you push it down with your finger), it sinks deeper into the water. If you remove the weight, the ice cube bounces back up — or rebounds.

IN THIS CHAPTER

» Describing regions where wind is important

» Moving sediments with wind currents

» Spotting features of wind erosion

» Depositing sediments as dunes and loess

» Debating the formation of desert pavement

Chapter **14**

Blowing in the Wind: Moving Sediments without Water

M uch of the earth's surface is subjected to change by forces of water (see Chapters 12, 13, and 15). But in regions that lack water, such as deserts, wind is the most important geologic force; it creates geologic features by removing rock particles *(erosion)* and adding new sediments *(deposition)*. The term *aeolian* (sometimes spelled *eolian*) describes the processes and features of wind.

In this chapter, I explain how wind shapes the earth's surface and describe geologic features associated with the processes of wind erosion and deposition.

Lacking Water: Arid Regions of the Earth

Wind creates prominent geologic features in areas of the earth that do not have much water. These relatively dry or *arid* regions occur anywhere on the planet that the amount of *precipitation* (rain and snowfall) is less than the amount of water lost to *evaporation* (the transformation of liquid water into water vapor by heat). When the sediments are dry, they become susceptible to transport by wind.

Unlike what you may see in movies and television, the arid regions of Earth are not all hot, sunny, sand-covered expanses. On the contrary, only a few arid regions meet that description, such as the Arabian Desert of Saudi Arabia. Other arid landscapes occur in cold climates (such as Antarctica), along coastlines (such as the Atacama Desert of South America's west coast), and in mountainous continental interiors (such as the Himalayas of Asia and the Rocky Mountains of North America).

When water is present in arid regions, the moving water in streams creates geologic features, but wind is still by far the most common mover of earth materials.

Transporting Particles by Air

Wind carries particles of rock, or *sediment,* in much the same way that moving water carries sediment. (I describe movement of sediments by water in Chapter 12.) Both wind and water flow in currents and pick up larger particles as their flow increases. The biggest difference between the two is the size of the particles that can be carried.

Air is much less *dense* than water, which means there is more space between the molecules in a given volume of air than between molecules in the same volume of water. (See Chapter 10 for more details on density.) The result of this extra space (and therefore lower density) is that wind must move much faster to pick up particles the same size that water picks up at lower speeds.

Like water, wind flows in two ways:

>> **Laminar flow:** Its lines of flow, or *streamlines,* are unmixed and proceed in straight lines.

>> **Turbulent flow:** The streamlines are mixed.

I show an illustration of laminar and turbulent flows in Chapter 12.

The methods of particle movement, or *transport,* by wind are the same as some of the methods described for particle movement by water in Chapter 12. I define them here briefly.

Skipping right along: Bed load and saltation

Particles transported by wind that move along very close to the ground are called wind's *bed load.* Most sediments transported by wind move along as bed load, close to and sometime in contact with the surface. This fact is due to the low density of air, which limits its ability to lift large particles, even at high speeds.

To move sand-sized particles, wind first pushes them along the surface. If the particle is too heavy to be lifted, it continues to move forward along the ground in a movement called *creep.*

Smaller or lighter sediments are picked up into the wind flow when the speed (or *velocity*) of the wind flow is great enough to lift them up. But even after they're lifted, the particles are still subjected to the pull of gravity, so after a short period of suspension they fall back to the surface.

This type of particle motion is called *saltation:* The particle jumps along the surface, being repeatedly lifted and dropped as it moves in response to the force of the wind flow.

As the particles land, they bounce farther along and bump other grains, which may begin to *saltate* or move forward in the same manner. In this way, the bed load of a wind current is filled with sand-sized sediments bouncing along in the direction of the wind flow. If the surface is very hard, the bed load may extend up to 2 meters (about 6.5 feet) above the surface because objects bounce higher off a more compact surface. (For example, if you drop a marble onto grass, it will not bounce or move very far. But if you drop the same marble onto pavement, it will bounce and move much farther.)

The bed load of a wind current is illustrated in Figure 14-1.

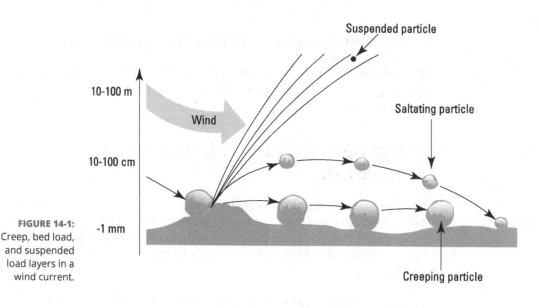

FIGURE 14-1:
Creep, bed load, and suspended load layers in a wind current.

Suspending particles in air

Particles that are lifted completely away from the surface and carried within the flow of the wind current are called a *suspended load.* Particles suspended by wind are usually silt or clay particles measuring less than 0.15 mm. These small particles (dust) can be picked up and lifted high into the atmosphere by turbulent wind flow and may remain suspended for years, traveling across great distances (thousands of kilometers) before settling back to the earth's surface.

A wind current could not have a suspended load of particles without a bed load. The speed of wind flow close to the surface is very low — so low that even the small particles of dust cannot be picked up without some other force instigating their movement. The saltation of larger particles provides this force: The bouncing of the larger grains unsettles the smaller grains, pushing them into the air flow where they can be lifted and suspended in the wind. This is also illustrated in Figure 14-1.

Deflating and Abrading: Features of Wind Erosion

As wind picks up and carries particles of sediment away, it creates features of aeolian erosion. Unlike water or ice, wind cannot remove sediment grains from solid rock. But in arid regions, there is less moisture and vegetation to anchor or hold the loose sediment particles to the surface. This means the loose particles that have been removed from rocks by a previous process of erosion or weathering (see Chapter 7) are available for movement by the wind.

Removing sediments

Wind *deflates* a surface, lowering it as sediments are removed. This process of wind erosion is called *deflation*.

The most common geologic features of deflation are *blowouts*. Many blowouts are small, shallow areas, sometimes only a few meters wide and less than a meter deep, where deflation has occurred. However, other blowouts, called *deflation basins*, are much larger — sometimes many kilometers across.

Blowouts occur in areas where vegetation has been removed, leaving loose sediments susceptible to wind erosion. Because particles must be dry for wind transport, the boundaries of blowout features are defined by the surrounding groundwater. When the blowout is deflated low enough that the sediments are moistened by groundwater, deflation stops because the particles can no longer be easily removed by wind.

Scratching the surface

After a wind current has picked up some sediment particles, it brushes these particles against other objects. You may have experienced this process if you've ever felt sand stinging your legs at the beach. The stinging feeling is the brushing of suspended and saltating sediment particles against your legs.

The blasting of sediments against another object by wind creates features of *wind abrasion*. Abrading by wind appears to polish some surfaces, making them a little shiny and very smooth.

Rocks that are subjected to wind abrasion have recognizable features and are called *ventifacts*. Ventifacts have multiple surfaces eroded by wind abrasion. Figure 14-2 illustrates how a ventifact is created.

First the wind abrades one side of the rock, creating a smooth, flat surface in the direction of wind flow. If the rock's position changes or the direction of the wind flow changes, another flat, smooth surface is created by abrasion.

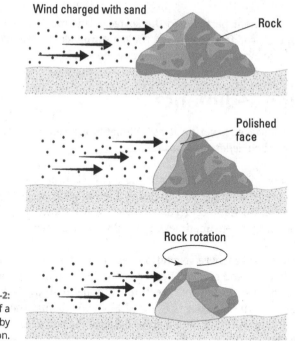

Wind charged with sand

Rock

Polished face

Rock rotation

FIGURE 14-2:
Creation of a
ventifact by
abrasion.

Just Add Wind: Dunes and Other Depositional Wind Features

Geologic features created by wind depositing (leaving behind) sediments are usually one of two types: *dunes* and *loess*. The distinguishing characteristic of these two features is the size of the particles that build them. Due to the limited ability of wind to keep particles lifted into its flow, significant *sorting* of sediments by size occurs as wind transports particles. Sorting means that sediments of similar size are grouped together when they are deposited.

Sand particles are the largest sediments moved by wind. As soon as the wind flow slows down, the sand particles fall to the surface, while the smaller particles of clay and silt remain carried as suspended load in the wind current. The result is that some features of wind deposition, the *dunes*, are created almost entirely of sand-sized particles. The smaller particles of silt are deposited when the wind stops moving completely, forming layers of *loess*.

In this section, I describe these distinct features and explain how they are created by wind deposition.

Migrating piles of sand: Dunes

Sand dunes form in areas where the wind blows almost constantly and there is a large supply of sand that is not held in place by vegetation or moisture (such as deserts or sandy coastlines). When the wind flow slows, the sand in its bed load is deposited in piles, creating dunes.

Sand dunes are usually asymmetric in appearance, with one side having a steeper slope than the other. Figure 14-3 illustrates the different parts of a typical sand dune.

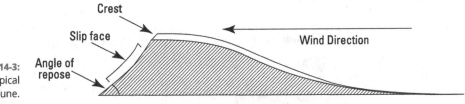

FIGURE 14-3:
A typical sand dune.

As wind moves sand up the *windward* side of a dune, most of the sand is deposited at the *crest* or top of the dune. When this pile of sand is too heavy to be supported by the steeper angle on the *leeward* side of the dune, the sediments slip downward in response to the pull of gravity, creating a *slip face.* The sediments rest along the slip face at the *angle of repose* (which I describe in detail in Chapter 12).

The sliding of sediments down the slip face of a dune create layers, as illustrated on the left in Figure 14-4. Over time, multiple dunes are deposited above, creating a series of *cross-beds* such as illustrated on the right in Figure 14-4. These cross-beds are preserved in the rocks formed by these sediments and are snapshots of dunes that existed when the sediments were first deposited.

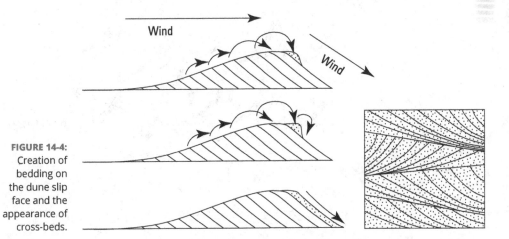

FIGURE 14-4:
Creation of bedding on the dune slip face and the appearance of cross-beds.

The ongoing process of wind moving sediments up the windward face of a dune by saltation, and down the leeward face of a dune by slipping, causes the dune to move across the surface in the direction the wind is blowing. This movement is called *dune migration*.

Shaping sand

You may think the formation of dunes by wind creates a never-ending array of different shapes. However, only a few major shapes or types of dunes are consistently created by wind. I describe them here, and they are illustrated in Figure 14-5.

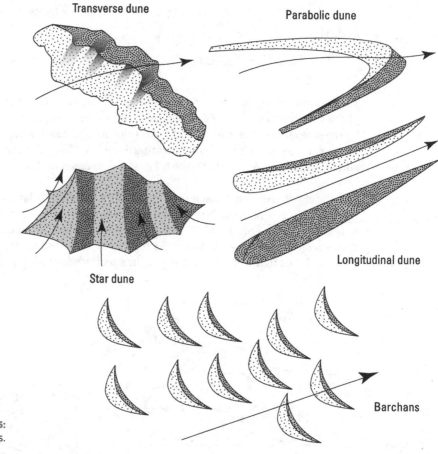

FIGURE 14-5:
Dune types.

Barchan dunes

Barchan dunes are created when the sand supply is limited and the land surface is flat. They are crescent-shaped, like a half moon, with two points or *horns* pointed *downwind,* or in the direction the wind is blowing. Most commonly the horns are the same length, but occasionally one is longer than the other.

Parabolic dunes

Parabolic dunes appear similar to barchan dunes in shape, except the horns of their crescent shape point *upwind,* in the direction the wind flows from, instead of downwind. Parabolic dunes are most common along coasts, where the sand is partially anchored by vegetation. The result is that sand is moved forward in areas without vegetation (the curve of the crescent) and remains in place where vegetation is present (at the horns).

Transverse dunes

In a region with high amounts of sand, transverse dunes create the appearance of waves across a sandy sea. Transverse dunes are ridges of dunes lined up perpendicular to, or crossing the direction of, wind flow. This type of dune is commonly found along the coast, on beaches where the wind blows steadily toward land and there is plenty of sand.

Longitudinal dunes

Longitudinal dunes also form in lines, similar to transverse dunes, but the lines of longitudinal dunes are parallel to (in the same direction of) wind flow. They most commonly occur where wind direction changes slightly within a narrow range.

Star dunes

Star dunes are complex dune shapes usually confined to large, sandy deserts such as the Sahara in Northern Africa. They are the result of wind blowing from many different directions (at least three) or shifting its direction constantly. A star dune features a high central point and numerous ridges of sand coming off it, creating a star shape.

Laying down layers of loess

If dunes result from wind depositing its bed load particles, what happens to all the smaller sediments carried as suspended load in the wind? The particles of silt and clay suspended in wind currents are deposited as *loess.*

LIKE DUST IN THE WIND: THE ECOLOGICAL THREAT OF DESERTIFICATION

Desertification is the process of changing productive farm or grazing land into barren desert. The loss of useful farmland to desertification is one of the major threats of climate change. This shift is devastating to farmers, ranchers, and communities that depend on the land for food and income.

The most common cause of desertification is allowing animals to *overgraze* or eat all the vegetation in a confined area. When the vegetation is gone, the topsoil — the most valuable nutrient-rich sediment — dries out and is blown away as dust in the wind. The presence of vegetation provides a physical anchor; the roots hold sediments in place. Vegetation also provides increased moisture to the soil. When the vegetation and topsoil are removed, the land becomes nearly useless: No crops grow, and no plants are available for animals to graze. The humans dependent on this land for food have to move somewhere else, where they may start the process all over again, spreading the desertification throughout an entire region.

The spread of desertification particularly threatens regions that are already arid and depend on the grazing of herd animals for survival, such as southern Africa and northern China. But desertification also occurs in semi-arid regions, including the Mediterranean and the American Great Plains. Many countries have begun to attack this problem by creating strategies for smart land use, including rotational grazing, as well as working to reduce the effects of greenhouse gas climate warming.

Here's the process: Small-sized sediments are picked up by wind in glacial or desert environments, carried far away, and deposited as loess whenever and wherever the wind stops moving, often blanketing whole regions. For example, much of the Midwestern United States and significant portions of the Pacific Northwest are covered in loess that was deposited thousands of years ago when large parts of North America were covered by ice sheets (see Chapter 13).

Loess sediments have some unique qualities that make them desirable as farmland:

>> **Loess particles have not been altered by chemical weathering.** The fact that loess sediments originate in dry environments (including glacial deposits) means that they have not been subjected to the heavy chemical weathering

(see Chapter 7) that removes important elements from the minerals. This means these elements are available in the loess deposits as mineral nutrients to support vegetation.

>> **Loess particles are windblown and angular.** Suspended particles carried by wind are not rounded like particles carried by water. (Think of how rounded river rocks are.) Rather, wind transport and abrasion create sharp, angular edges on the particles (similar to the ventifact, illustrated in Figure 14-2). Because of the angular shapes, when the particles are deposited they don't pack together very tightly, which means there is space between them for water to accumulate. The result is that water is available as moisture for plant roots.

Paving the Desert: Deposition or Erosion?

A unique feature found in arid environments that are subjected to processes of wind erosion and deposition is *desert pavement*. Desert pavement is a thin layer of pebbles and rocks across the land surface.

For a long time, geologists thought desert pavement was the result of erosion — the result of wind removing smaller particles from the surface, leaving behind a layer of larger pebbles and gravel. This layer left behind is called a *lag deposit* or *lag feature*. The process of creating desert pavement with erosion is shown in Figure 14-6.

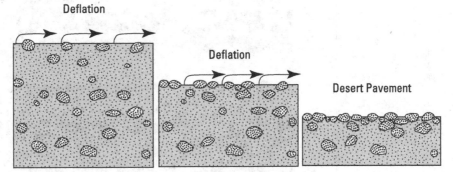

FIGURE 14-6: Desert pavement created by wind erosion.

However, some evidence suggests that desert pavement may be created by the deposition of clay sediments instead of their erosion. By this process (shown in Figure 14-7), clay and silt particles are deposited among the rocky surface materials and accumulate through time, creating a layer beneath the surface rocks.

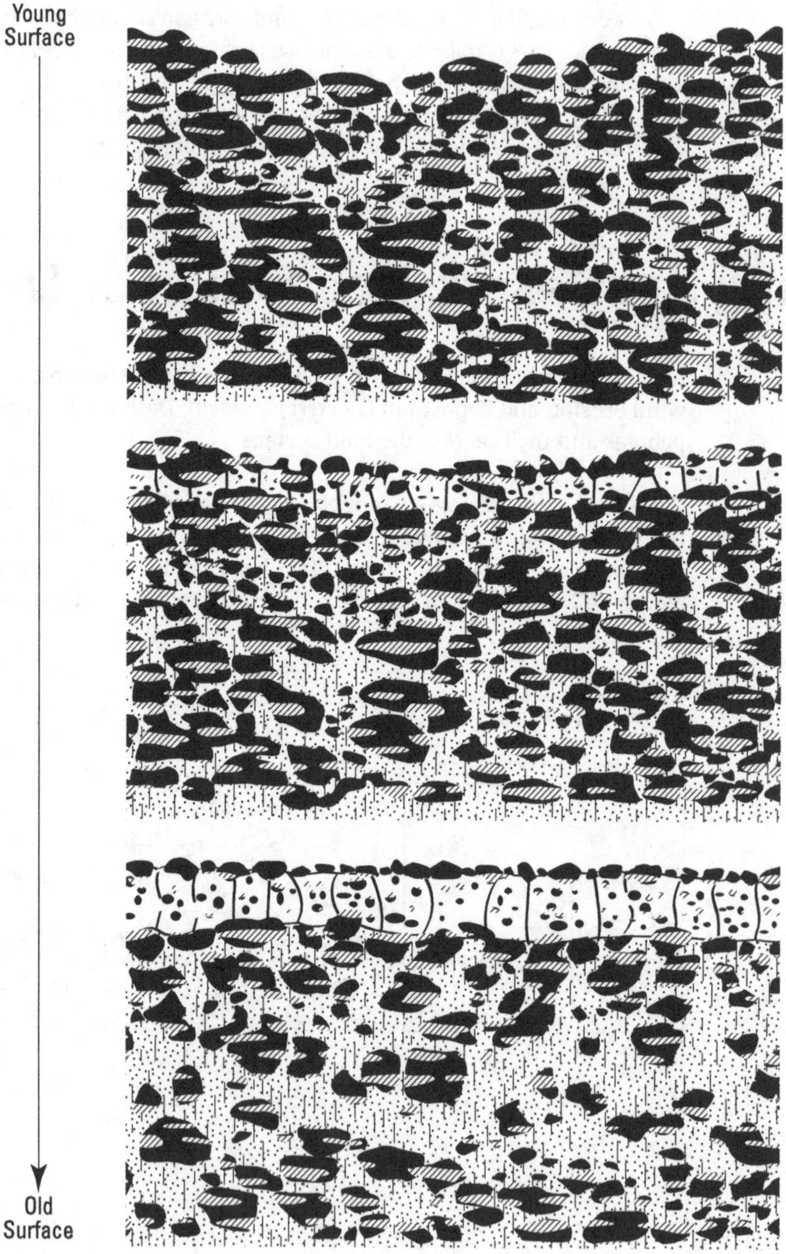

Young
Surface

FIGURE 14-7:
The formation of
desert pavement
through
deposition.

Old
Surface

Chapter **15**

Catch a Wave: The Evolution of Shorelines

A long the edges of continents, rocks meet water. In most of those places, called *coasts*, the water of the ocean is moving and through its movement changes the shape and characteristics of the shoreline.

In this chapter, I describe waves, wave motion, and the features created by the erosion and deposition of sediments along continental coastlines.

Breaking Free: Waves and Wave Motion

In order to accurately describe, record, and study ocean waves, scientists have defined and named different parts of a wave of water. In this section, I explain the parts of a wave and describe how waves roll across the ocean toward shore.

Dissecting wave anatomy

Waves are created when wind blows across the surface of water, pushing the water up into *crests*. Figure 15-1 illustrates the crest of a wave, as well as the low *troughs* in between each crest. These features are used to measure the wave height and length and to calculate its period and velocity. These wave features are defined here and illustrated in Figure 15-1.

>> **Height:** Wave *height* is the distance from the top of a crest to the bottom of the trough.

>> **Length:** Wave *length* is the distance from one crest to another or from one trough to another.

>> **Period:** The *period* of a wave is the time it takes a wave to pass a certain point.

>> **Velocity:** The *velocity* of a wave is the speed a wave is traveling.

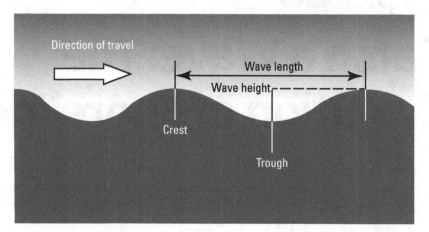

FIGURE 15-1:
Parts of a wave.

These features of a wave are determined by four characteristics of the wind:

>> The wind speed

>> How long the wind blows across the water surface, or its *duration*

>> The direction of the wind and whether it blows from one direction or multiple directions

>> The distance the wind moves across the surface of the water, or its *fetch*

Starting to roll

The force of the blowing wind causes the water to move in a rolling pattern. This pattern is called *oscillatory motion*. In the middle of the ocean, when the water moves in oscillatory motion, the water itself moves only a small distance while the energy of the wave keeps moving. This process is illustrated in Figure 15-2.

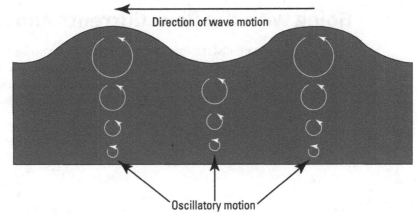

FIGURE 15-2:
Oscillatory
wave motion.

REMEMBER

This oscillatory motion is what causes boats to bob up and down in the ocean, moving only slightly in the direction of the waves. The circular motion of the water in oscillatory waves continues to circulate beneath the surface of the water to a depth approximately equal to half the distance of the wave's length.

When oscillatory waves move into shallow water, such as near the shore, the below-surface oscillatory wave motion is interrupted by hitting the bottom. In response to this friction, the wave slows down. Other incoming waves catch up to it, and they get bunched together in a traffic jam of waves approaching the shore. The combined water of the multiple waves builds a higher wave that is eventually too tall to support itself and collapses, or breaks, as it hits the shore; see Figure 15-3.

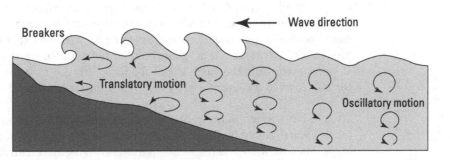

FIGURE 15-3:
Transition of
waves from
oscillatory to
translatory
motion in
shallow water.

These *breakers,* or waves that collapse and hit the shore, are waves of translation or *translatory waves.* Unlike the oscillatory waves of the open sea, translatory waves move the water some distance up onto the beach. The water moving up onto the beach is called *swash,* and the water that moves back toward the sea is called *backwash.*

Going with the flow: Currents and tides

Ruled by forces more distant than the wind, *tides* move ocean water toward and away from land in response to the pull of the moon's gravity. As the moon orbits the earth, its gravity pulls on the water at Earth's surface just slightly. The result of the moon's pull is that the water on Earth bulges out toward the moon, as illustrated in Figure 15-4.

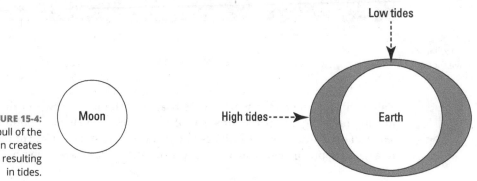

FIGURE 15-4: The pull of the moon creates a bulge, resulting in tides.

REMEMBER

The earth continues to rotate while the water bulges out, resulting in two high tides and two low tides every day. The *high tides* occur in the places that are closest to and farthest from the moon, where the bulging of water is the greatest.

Standing on the beach, the experience of tides is less abstract: *Low tide* is when the water is moving away from the land, and *high tide* is when it moves toward land. The flow of water with the tides creates *tidal currents*. When the tides are moving from low tide to high tide, the tide is "coming in" and the tidal current is called a *flood current*. When the tide is "going out," or switching to low tide, the movement of water away from the shore is called an *ebb current*. The areas along the shore that are covered and uncovered by the cycle of tidal currents are called *tidal flats*.

While tidal currents move in and out with the tides, other types of currents are moving water along the shoreline with the waves. When waves hit the beach at an angle, their motion creates *longshore currents*. Longshore currents move water (and sediments) parallel to the shore. A longshore current is generated by the energy of breaking waves moving back out and hitting incoming waves. The water is then forced to move along parallel to the shore. This process is illustrated in Figure 15-5.

If you have ever swum in the ocean and noticed that after a while you have moved along the beach away from where you left your towel, you have been moving with the longshore currents. This movement you experience is called *shoredrift*.

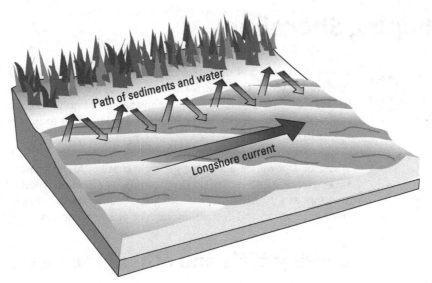

FIGURE 15-5:
Generating a
longshore
current.

Path of sediments and water

Longshore current

Another type of ocean current is a *rip current*. Rip currents occur when wave energy hits the shore straight on, and the returning water washes straight out to deeper water rather than being moved along the shore. This motion is illustrated in Figure 15-6.

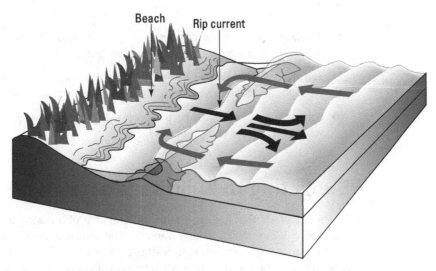

Beach   Rip current

FIGURE 15-6:
Motion of
a rip current.

WARNING

Rip currents are informally called "undertow," and when rip currents are especially strong, public beaches may be closed to prevent swimmers from being pulled out into dangerous water farther from shore.

CHAPTER 15 **Catch a Wave: The Evolution of Shorelines** 227

Shaping Shorelines

TIP

Waves, currents, and tidal motion change the shape of the coastlines by creating (and destroying) geologic features. In Chapter 12, I explain the relationship between the speed of flowing water and sediment grain size that determines how sediments are moved by water. The same rules apply to how water carries sediments in the oceans.

Water is a powerful agent for geologic change. Along coasts where it contacts rocks and cliffs, it erodes these materials — removing sediment particles small amounts at a time and transporting the sediments to other locations. The processes of erosion and deposition produce distinct features along the coast.

Carving cliffs and other features

Where waves meet rocky, steep shores they carve away sediments, creating a *wave-cut cliff*. Materials are removed from the cliff by the force of water, leaving behind a flat area below the cliff called the *wave-cut platform*. A rocky coastline being eroded will often have cliffs that extend out into the water, called *headlands*. These headlands are made of rock or sediment that resist erosion by the waves. Other features found along rocky coastal cliffs are illustrated in Figure 15-7 and include:

>> **Sea caves:** Sea caves are created as the waves remove easily eroded sediments and rock from the cliff, carving out a cave.

>> **Sea arch:** When two caves being carved out near each other connect (often on opposite sides of a headland), they form a sea arch.

>> **Sea stack:** After enough time, the top of the sea arch will weaken and fall, leaving tall, freestanding sea stack features.

Budgeting to build sandbars

All the materials that are removed from one area of a coast must end up somewhere else. Waves and currents transport the sediments and then deposit them, creating shoreline depositional features. Depositional features of a shoreline depend largely on the amount of sediments available to be transported and deposited, or the *sediment budget*.

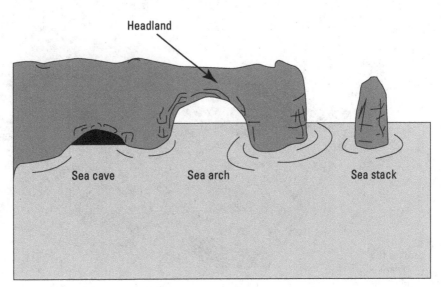

Headland

Sea cave   Sea arch   Sea stack

FIGURE 15-7:
Coastal features
of erosion.

REMEMBER

If a shoreline has a balanced sediment budget, the amount of sediment being eroded and carried away is being resupplied by sediment input so that the amount of sediment in the coastal environment remains nearly constant. Sources of sediment input include coastal erosion of headlands, as well as river deltas that may bring sediments from the interior of the continent.

Beaches are the most common feature of coastal deposition. A *beach* is the relatively narrow piece of land where the movement of currents, tides, and waves moves sediment particles. Some beaches are full of white sand, but a beach can have larger rocks, or rocks and sand of different colors, depending on what the source of the sediment is. For example, in the Hawaiian Islands, most of the rocks being eroded by waves are basalt formed from the eruption of Hawaiian volcanoes (see Chapter 7). This process results in dark (sometimes even black) sand beaches.

Other depositional features of shoreline processes are illustrated in Figure 15-8 and include:

>> **Spits:** *Spits* are long ridges of sand that project out into the water, often in a hook-like shape. Spits usually form parallel to the shoreline.

>> **Baymouth bar:** A spit that continues to grow until it closes a bay off from the ocean is called a *baymouth bar.*

>> **Tombolo:** A *tombolo* is a shallow ridge of sand that connects a small island or sea stack to the mainland along the coast.

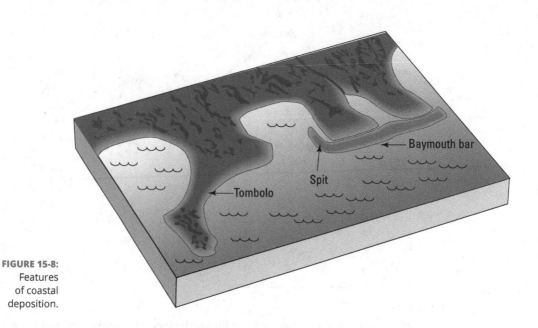

FIGURE 15-8:
Features
of coastal
deposition.

Baymouth bar

Spit

Tombolo

Categorizing Coastlines

Coastlines shaped by the processes of wave erosion and transport are called *secondary coastlines.* For example, much of the east coast of the United States is secondary coastline, shaped by marine processes.

Other coastlines, called *primary coastlines,* have features dominantly shaped by different surface processes (such as those discussed in Chapters 12 and 13). An example of a primary coast is a region where glaciers meet the ocean, such as the west coast of Canada and southeast Alaska. The carving of the landscape by the glacial ice (I describe glacial erosion in Chapter 13) is what creates the features of these coastlines.

Similarly, the coastal features of the northeastern United States (such as Cape Cod, Massachusetts) are the result of glacial deposition during the Pleistocene ice age. (Check out Chapter 21 for the scoop on recent ice ages.) The glacial features, including moraines (see Chapter 13), are still the dominant feature shaping the coastline.

Another type of primary coast occurs when large rivers empty into the sea. The processes of stream erosion and deposition associated with flowing water (I explain these in Chapter 12) shape the coastline in these regions, such as the Mississippi River outlet into the Gulf of Mexico in Louisiana.

HEADING FOR HIGHER GROUND: TSUNAMIS

A *tsunami* is a series of very large waves that travel across the ocean following an under-sea geologic event such as an earthquake, volcanic eruption, or landslide. Most often tsunamis are the result of undersea earthquakes. As illustrated in the first figure in this sidebar, the movement of the seafloor when an earthquake occurs displaces a large amount of water. This water then moves across the sea as a large wave.

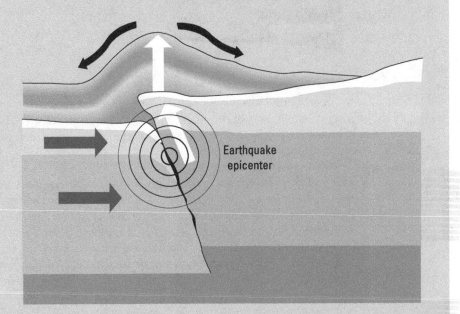

The danger in tsunami waves is that they are not visible the way normal ocean waves are, and they result in much more destruction when they reach the shore. In the deep ocean, a tsunami wavelength can be hundreds of miles in length and only a few feet in height, which means a boat floating in the ocean wouldn't even notice as the tsunami passed it. The wave can travel hundreds of miles an hour, crossing the entire Pacific Ocean in a matter of hours. This is comparable to the speed of a commercial airliner. In 1960, an earthquake on the coast of Chile in western South America generated a tsunami that hit the east coast of Japan less than 24 hours later.

The water in tsunami waves doesn't move in the circular, oscillatory motion that regular ocean waves exhibit, so when they reach the shore the tsunami waves don't build up into rolling breakers but proceed up onto the shore as illustrated in the second figure

(continued)

(continued)

in this sidebar. As the water gets shallower near shore, the wave height increases; a wave that is only a few feet tall in the open sea can build up to a more than 20-foot-tall wall of water as it moves onto the beach.

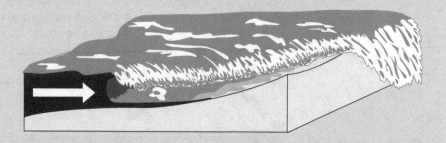

The force of all this wave energy pushes a large amount of water up onto shore, well past the normal sea level. (The height of water above normal sea level is called *runup*.) The water can proceed for miles inland, flowing for 30 minutes or more before gradually running out of energy.

After the Indian Ocean tsunami of 2004, research and public interest in understanding these destructive waves reached a peak. While the Pacific Ocean had experienced tsunamis in the past, the Indian Ocean tsunami of 2004 caught the attention of the entire world. The result is that scientists were funded to further study evidence of past tsunamis in an attempt to develop ways of predicating future tsunamis. The best bet is to prepare thorough advance warnings communicated throughout the international community, such as the one that constitutes the Tsunami Warning System in the Pacific, centered out of Hawaii. This system spreads messages and information through the region when geologic events that may trigger a tsunami occur.

Unfortunately, warning systems are most helpful to regions far away from the origin of the tsunami. In the 2011 tsunami that struck northern Japan, there was little time to warn people in the regions where the tsunami struck first (and most catastrophically). Regions across the Pacific Ocean including Hawaii, California, and South America, had plenty of warning and time to prepare and did not experience the violence and destruction of the early waves that struck the northern islands of Japan.

Not all coastal regions are susceptible to tsunamis — only the ones that are near regions of regular or intense earthquake activity. A few warning signs can help you recognize whether a tsunami may be headed your way. If you are near the coast and you feel an earthquake; see the water rapidly move out away from the shore like an extra-low tide (it often does this before the rush of tsunami water hits the shore); and hear a loud, train-like noise, the best thing to do is head for higher ground.

5

Long, Long Ago in This Galaxy Right Here

Discover how scientists determine the age of rocks using radioactivity and *stratigraphic* relationships (the relationships among various layers of rock).

Find out how Earth's history is organized into eons, eras, and periods using the geologic time scale.

Go layer by layer through the story of Earth, describing the geologic events and the evolution of organisms from single cells, to dinosaurs, to you and me.

Chapter **16**

Getting a Grip on Geologic Time

E ach time a rock forms, it preserves a snapshot of the earth processes that create it. This means that within the rocks of Earth's crust is a story of all the past geologic events in Earth's history. However, with the constant movement of plates (as described in Chapter 9) and changes to the earth's surface (described in Part 4), the history of events gets all mixed up. The challenge for geologists is to place the events in the right order and interpret the stories told in the rocks.

Geologists have created an accurate sequence for past geologic events by combining methods of relative dating with modern techniques of absolute dating. By combining the two, they constructed — and continue to revise — Earth's geologic timescale. The geologic timescale of Earth's history documents the past 4.5 billion years, beginning with Earth's formation (see Chapter 3). As new information is uncovered, the timescale is revised, updated, and improved, and it becomes more accurate.

In this chapter, I describe relative and absolute dating methods, explain how they are used to interpret Earth's past, and discuss how new research continues to refine and revise the details of Earth's history.

The Layer Cake of Time: Stratigraphy and Relative Dating

Organizing rock layers into the order of their formation is the first step in constructing a geologic history. When the events are in the proper order, the story in the rock layers can be read. The challenge is figuring out the original order of rock layers that have been tilted, deformed, eroded, and otherwise changed.

In this section, I describe how the study of rock layers, called *stratigraphy,* is used to untangle sequences of rock formation and construct a relative timescale for these events.

Speaking relatively

One way of describing history is to figure out the correct order of past events in relation to one another. This approach is called *relative dating.* Relative dating does not provide numerical ages or dates for events. Relative dating techniques describe only when events happen in relation to other events.

For example, if you describe the ages of your family members relative to you, you may say that your sister is younger than you are and your brother is older than you are. You describe each person's birth in relation to your birth: Your brother was born, then you were born, and then your sister was born. In this way, you construct a relative sequence of events without providing the actual age of each family member.

This same approach is used by geologists to build a relative history of the earth, stating that one period or sequence of events occurred before or after another.

Sorting out the strata

Long ago, geologists realized that rock layers could be organized through time relative to one another. This sequence of relative rock layers provided a context to interpret Earth's history. For this reason, the study of layered rocks is the foundation of relative dating in geology.

The study and interpretation of rock layers (or *strata*) is called *stratigraphy.* Scientists called *stratigraphers* compile information from rock layers all around the world and compare that information in order to understand the sequence of rock layer formation and details about the history of Earth's surface.

Stratigraphers are interested in both the history of the rock layers and their formation or origin. Therefore, these scientists have multiple ways of describing the rocks layers, or *units*:

» **Lithostratigraphic units:** When stratigraphers describe rock layers only by their physical characteristics, such as rock type, they use lithostratigraphic units. The basic lithostratigraphic unit is the *formation,* which can be separated into smaller units called *beds* or grouped into larger units called *groups* or *supergroups.*

» **Biostratigraphic units:** *Biostratigraphic units* are rock layers described by the fossils they contain, with no consideration of rock type.

» **Chronostratigraphic units:** *Chronostratigraphic units,* or time-stratigraphic units, are rock layers created during a specific time interval. Rocks in a chronostratigraphic unit may be of different types and have different types of fossils in them.

Putting rock layers in the right order

As I describe in Chapter 7, sedimentary rocks are formed in layers as sediment particles are deposited atop one another. Because these rocks and layers are subject to physical laws of gravity, certain assumptions can be made about how the layers in sedimentary rocks are created.

The first laws or principles of stratigraphy were described by the father of modern stratigraphy, Nicholas Steno (see Chapter 3). The principles of stratigraphy provide a way for geologists to put rock strata in the correct order of their formation. Doing so is especially useful when looking at rocks that have been folded or faulted (see Chapter 9) or have otherwise changed position since they formed. In Chapter 3, I explain Steno's four principles of stratigraphic correlation, or relationship. Here's the quick rundown:

» **Principle of superposition:** In rocks that have been undisturbed since they were formed, the layers of rock below are older than layers of rock above.

» **Principle of original horizontality:** All sedimentary rock layers are originally created in a horizontal position. If they are now vertical, they have been shifted after they were formed.

» **Principle of lateral continuity:** As rock layers form, each layer continues across a horizontal surface until another object or geologic feature stops it.

» **Principle of cross-cutting relationships:** When one rock type cuts through another, the cutting rock is younger than the rock it cuts through.

The first three of the preceding principles apply only to the sedimentary rocks, whereas the fourth one helps determine the age of igneous rocks in relation to the sedimentary layers they cut across as dikes and sills (see Chapter 7).

Two other principles are also used in modern stratigraphy:

>> **Principle of fossil succession:** *Fossils* (the remains of previous organisms) occur in a determinable order, and any time period can be recognized by the fossils present in rocks formed during that time. This principle applies only to sedimentary rocks because they are the only ones that can contain fossils.

>> **Principle of inclusions:** If a rock contains pieces of a different rock (called *inclusions*), the pieces must be from an older rock. This principle is especially useful for sorting out a sequence of events that includes igneous or metamorphic rocks, along with sedimentary ones.

Losing time in the layers

Together, the six principles of stratigraphy provide scientists the ground rules for putting rock layers in their correct order for interpretation of the past. However, each principle is based on the assumption that the rocks you examine have an undisturbed and complete record — that the sequence of rock layers is *conformable*.

Very often, pages of time are missing in the geologic record. These breaks in the rock record are called *unconformities*. An unconformity occurs when rock layers have been eroded, moved, or otherwise changed, and then more rock layers are added above. This process results in a period of time for which no rock record exists.

Fortunately, when periods of time are missing from the rock layers, certain clues signal this fact to the stratigrapher. There are three different types of unconformities, and each offers its own clues:

>> **Angular unconformity:** This unconformity occurs when rock layers are tilted at an angle by uplift, faulting, or folding (see Chapter 9); the layers are eroded; and then new horizontal layers are created above them. This process is illustrated in Figure 16-1.

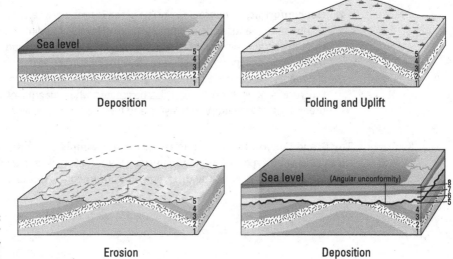

Deposition

Folding and Uplift

FIGURE 16-1:
How an angular
unconformity
is created.

Erosion

Deposition

>> **Disconformity:** A disconformity is the most difficult unconformity to recog-
nize. This is because both layers are sedimentary rock, and both are still
horizontal, but they are separated by a surface that was eroded, removing
part of the rock record. The older rock layers are still positioned horizontally
when the new rock layers form above them, as illustrated in Figure 16-2.
A stratigrapher may not recognize the disconformity in the rock sequence
until he or she has applied other stratigraphic principles (such as the principle
of fossil succession) that indicate time is missing.

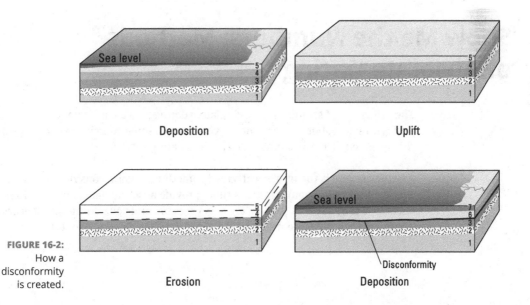

Deposition

Uplift

FIGURE 16-2:
How a
disconformity
is created.

Erosion

Deposition

>> **Nonconformity:** A *nonconformity* is a break in time between a metamorphic or intrusive igneous rock (see Chapter 7) and layers of sedimentary rocks created above it. A nonconformity is created when underground rocks (the metamorphic or igneous rocks) are lifted to the surface and eroded, and then sediments are deposited on top of them, which eventually become sedimentary rock. A nonconformity can also be created when magma intrudes into preexisting sedimentary rock layers and forms an intrusive igneous rock.

The Grand Canyon in Arizona is an excellent example of stratigraphy exhibiting these different types of unconformities. Figure 16-3 is a sketch of the Grand Canyon stratigraphy with the different types of unconformities labeled.

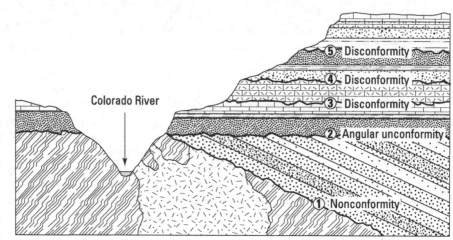

FIGURE 16-3:
The Grand Canyon exhibits a nonconformity (1), an angular unconformity (2), and multiple disconformities (3, 4, and 5).

Show Me the Numbers: Methods of Absolute Dating

The principles of stratigraphy and relative dating provide a sequence of events for geologic history, but they cannot tell you how old something is in years. To answer that question, scientists must use absolute dating techniques.

An *absolute date* is the numerical age of something. Again, consider describing the age of your family members. This time, provide absolute dates instead of relative dates. You may say that your sister is 15 years old and your brother is 27 years old. These are their absolute, numerical ages — nothing relative about it!

Measuring radioactive decay

The most common method used to determine the numeric age of rocks is called *radiometric dating.* This method measures the *decay,* or atomic changes, in certain atoms. To help you understand radioactive decay, I need to review briefly some information about atoms from Chapter 5.

Each atom has a nucleus with both protons and neutrons and is surrounded by electrons. The number of protons and neutrons in the nucleus determines each element's *atomic mass number.* The number of protons is the *atomic number.*

Some elements, such as carbon, may have atoms with different numbers of neutrons in their nucleus. This deviation changes the atomic mass number but not the atomic number. Atoms that contain the same number of protons but different numbers of neutrons are called *isotopes.*

Isotope names are written with the element and the atomic mass number. For example, carbon-13 and carbon-14 isotopes exist, as does normal, nonradioactive carbon — carbon-12 — which has an atomic mass number of 12. (You may sometimes see these isotopes written as ^{12}C, ^{13}C, and ^{14}C). Each type of carbon has the same number of protons (6) but different numbers of neutrons (6 neutrons in carbon-12, 7 neutrons in carbon-13, and 8 neutrons in carbon-14).

REMEMBER

Some isotopes are unstable, or *radioactive,* which means changes automatically occur within the nucleus that transform the isotope into a completely different element by changing its atomic number. As I explain next, this radioactive decay of certain isotopes is a useful tool for absolute dating.

REMEMBER

If the number of neutrons in an atom of an element changes, the atomic mass number changes, which creates an isotope. If the number of protons in an atom changes, both the atomic number and the atomic mass number change, and you have a completely different element.

Three ways to decay

Radioactive elements transform in one of three ways:

>> **Alpha decay:** *Alpha decay* describes the change to an atom when two protons and two neutrons leave the nucleus. The result is an atom of a different element that has a different atomic mass number. For example, when the isotope of uranium (which has 92 protons in its nucleus and 146 neutrons) with an atomic mass number of 238 (uranium-238) experiences alpha decay, the result is an atom of thorium-234. Thorium-234 is an isotope with an atomic mass number 234 and an atomic number of 90. Alpha decay is illustrated in Figure 16-4.

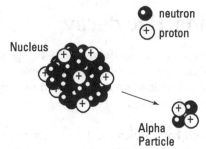

FIGURE 16-4:
Alpha decay of
a radioactive
isotope.

neutron
proton

Nucleus

Alpha
Particle

>> **Beta decay:** *Beta decay* describes when a neutron splits itself into two
separate particles — an electron and a proton — and the electron leaves
that atom. The result is that there is an additional proton in the nucleus, which
changes the atomic number and transforms the atom into a new element. For
example, the isotope potassium-40 (with an atomic number of 19) becomes
calcium-40 (with an atomic number of 20). The atomic mass number of these
two atoms is the same, 40, because one neutron was replaced by one proton
in the nucleus by the process of beta decay. Beta decay is illustrated in
Figure 16-5.

Beta decay

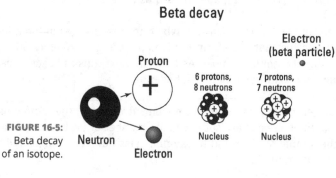

Electron
(beta particle)

Proton

6 protons,
8 neutrons

7 protons,
7 neutrons

FIGURE 16-5:
Beta decay
of an isotope.

Neutron

Electron

Nucleus

Nucleus

>> **Beta capture:** *Beta capture* is the reverse of beta decay. In beta capture, an
atom grabs an electron from somewhere else and combines it with a proton to
create a neutron in its nucleus. The result is a change in the atomic number but
not the atomic mass number. For example, when potassium-40 (with an atomic
number of 19) experiences beta capture, it becomes argon-40 with an atomic
number of 18 because one proton from the nucleus of the potassium-40 atom
has been combined with the electron to create the new neutron. Beta capture is
illustrated in Figure 16-6.

Beta capture

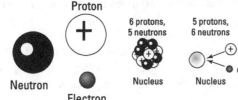

FIGURE 16-6:
Beta capture of an isotope.

Neutron

Proton

Electron

6 protons, 5 neutrons

Nucleus

5 protons, 6 neutrons

Nucleus

p^+

e^-

REMEMBER

Keep in mind that not all elements have isotopes and that not all isotopes are radioactive. Only unstable isotopes are radioactive, and after they transform into stable isotopes of another element, they are no longer radioactive.

Fortunately for scientists, the transformations of radioactive isotopes occur in such a way that they can be measured and used to date the formation of minerals.

It takes a half-life: Transforming parent isotopes to daughter isotopes

Radioactive isotopes begin as one element and, through decay, become another. The original element is called the *parent isotope,* and the resulting element is called the *daughter isotope.* Comparing the number of parent and daughter isotopes and knowing how long it takes for each radioactive element to decay is what provides a numerical age for rocks.

When a rock forms from cooling magma (see Chapter 7), minerals are formed that incorporate into their crystal structure whichever elements are available. (I explain minerals' crystal structure in Chapter 5.) For example, a mineral of biotite will use potassium atoms in its crystals. If some radioactive isotopes of potassium are in the magma, then some of the potassium atoms that form biotite minerals from that magma will be radioactive. Over time, the unstable potassium isotopes in the biotite mineral decay, transforming into argon atoms, but they remain part of the biotite mineral structure.

REMEMBER

The decay of a radioactive isotope can be measured in half-lives. The *half-life* is the amount of time it takes for half the parent isotope atoms to decay into daughter isotope atoms.

Scientists have discovered that decay rates of radioactive isotopes are constant: They do not change in response to temperature, pressure, or other chemical factors. Therefore, when the decay rate for a radioactive isotope is known, measuring the number of parent and daughter atoms in a rock or mineral can be used to determine, in years, how many half-lives have occurred since the mineral (or rock) was created.

For example, if a rock forms with 100 radioactive parent isotopes, after one half-life the rock will have 50 parent isotopes and 50 daughter isotopes. After two half-lives it will have 25 parent isotopes and 75 daughter isotopes.

Because half of the existing parent isotopes decay in each half-life period, there is a predictable ratio of the percentage of parent-daughter atoms that tells you how many half-lives have passed:

>> When the rock is formed, it always has 100 percent of the parent isotope.

>> After one half-life, 50 percent of the parent isotope remains.

>> After two half-lives, 25 percent of the parent remains.

>> After three half-lives, 12.5 percent of the parent remains.

And so on until the amount is too small to measure or no parent atoms remain.

Common radioactive isotopes for geological dating

Not all radioactive isotopes are useful for dating. Some of them have half-lives that are much too long or much too short for humans to measure. A good half-life range for dating earth materials is somewhere around several thousand years.

TIP

Choosing the best isotope to date a particular rock depends on how old you think the rock may be, which minerals (and therefore elements) are in the rock, and which type of rock you are dating. Table 16-1 lists and describes the most common radioactive parent-daughter isotope pairs that are used in dating rocks.

The long half-lives of many of these methods allow geologists to date some of the most ancient rocks on Earth (and the moon!). How does a geologist know which method to choose if he doesn't know the age of the rocks? First he must use relative dating and stratigraphy to estimate how old the rocks are, and then he can choose the best radioactive isotope system that is present in the minerals of his rock. Even so, using more than one isotope pair measurement can provide the best result.

TABLE 16-1
Commonly Used Radioactive Isotopes for Dating

Dating Method Name	Parent	Daughter	Half-Life	Effective Range for Dating (Years)	Materials Dated
Rubidium-strontium	Rb-87	Sr-87	47 billion years	10 million–4.6 billion	Biotite, muscovite, or potassium feldspar minerals in very old metamorphic or igneous rocks
Uranium-lead	U-238	Pb-206	4.5 billion years	10 million–4.6 billion	Zircon or uraninite crystals in igneous and metamorphic rocks
Uranium-lead	U-235	Pb-207	713 million years	10 million–4.6 billion	Zircon or uraninite crystals in igneous and metamorphic rocks
Thorium-lead	Th-232	Pb-208	14.1 billion years	10 million–4.6 billion	Zircon or uraninite crystals
Potassium-argon	K-40	Ar-40	1.3 billion years	100,000–4.6 billion	Volcanic rocks, muscovite, biotite, horneblende minerals
Carbon-14	C-14	N-14	5,730 years	Within the last 60,000 years	Any once-living organism

REMEMBER

When two isotope pairs are measured and they result in the same (or very similar) age for a rock, the dates are described as *concordant* or in agreement. If the two dates are not very similar, the dates are called *discordant,* and the scientist must take new samples and new measurements until there is enough concordant result to draw a trustworthy conclusion.

Other exacting methods of geological dating

Radiometric dating is by far the most common method of assigning numerical ages to rocks and rock layers, but other methods provide absolute ages as well.

Fission-track dating

Fission-track dating is a method of counting the number of fission tracks in a mineral. These are very small trails etched in a crystal when radioactive atoms of uranium decay. As long as the minerals have not been exposed to high temperatures, the fission tracks will be present. Scientists can count the fission tracks and calculate an age for the rock. But if the minerals have been exposed to high temperatures, the trails will be erased, and the calculations of any visible trails will provide an age younger than the actual age of the rock.

EXPOSING COSMIC RADIATION

The age of rock and mineral formations is only one of the questions earth scientists have about past geologic events. Many researchers want to know not only when the rock was formed but also when the rock reached the surface of the earth, or became exposed. Answering this question may provide geologists with a way to document the age and sequence of past dramatic geologic events such as earthquakes, faulting, and uplift (see Chapter 9).

The answer is found in cutting-edge research into intergalactic cosmic rays and isotopes called *cosmogenic nuclides*. Similar to radioactive isotopes, cosmogenic nuclides (or cosmogenic isotopes) decay only when they have been exposed to cosmic rays. These "rays" are streams of high energy particles that pass through Earth's atmosphere and into the first 3 meters (10 feet) or so of the earth's surface. (You probably haven't noticed, but they are passing through you right now!) As these particles collide with atoms in the rocks of Earth's crust, they knock other particles out of the way — creating cosmogenic isotopes.

Unlike radioactive isotopes, which have a fixed number of parent atoms, cosmogenic isotopes are constantly being produced as the cosmic ray particles hit Earth's surface. Therefore, in order to use these isotopes as a method of dating, scientists must study and record their production rates as well as their half-lives.

Because cosmic rays can't affect atoms in a rock until the rock is at the surface of the earth, measuring the cosmogenic isotopes provides an age for when that rock became exposed. This tool is used by some geologists to study rates of uplift and erosion.

COUNTING TREE RINGS

Another method of absolute dating that is useful for very recent history is a method called *dendrochronology*. Dendrochronology examines the growth rings in trees and uses them to count back in time year by year. How is this possible? Each year, a tree produces a new ring of wood around its trunk. If the environmental conditions are particularly good, such as when high amounts of rain fall, the ring will be thicker than in years when conditions are not so good.

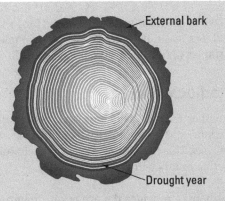

External bark

Drought year

With these changes visible in the patterns of the rings, scientists called *dendrochronologists* can compare the rings of multiple trees — a method called *cross-dating* — to build a bridge into the past.

Dendrochronology has been most useful in archaeological and climate studies of the American southwest. In regions that are arid, trees live a long time and show obvious changes to their growth ring pattern in years of drought or extra rainfall. By correlating these patterns with historic records of weather, accurate calendar years can be assigned to each ring.

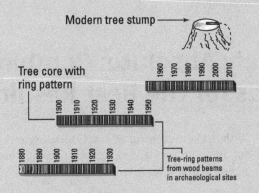

Modern tree stump →

Tree core with ring pattern

1960 1970 1980 1990 2000 2010

1900 1910 1920 1930 1940 1950

1880 1890 1900 1910 1920 1930

Tree-ring patterns from wood beams in archaeological sites

Hundreds of years ago, ancient societies living in the American southwest used wood beams to construct their homes. These beams have been preserved, along with other remnants of their culture and society. The beams provide a long record of tree rings for cross-dating and examining not only the history of people in the southwest but also the history of climate change (because the trees record years of drought and changes in rainfall).

TIP

Fission-track dating is most useful for rocks between 50,000 and 1.5 million years old, but it can be used for younger samples (a few hundred years old) and older samples (a few hundred million years old).

Radiocarbon

Another radioactive isotope that decays with a known half-life is the isotope of carbon-14. Maybe you've heard of *radiocarbon* or *C-14 dating* in television shows or movies. This method, used also by archaeologists, is the most common method of dating events within human history.

Carbon-14 provides the age of any material that used to be living and, therefore, has carbon in it. This includes bone, wood, shells, and paper. While carbon can be found in some minerals, it is most common in living organisms, who take in carbon atoms from the atmosphere while they are living. After the organism dies, the carbon-14 isotopes begin to decay.

TIP

The half-life of C-14 is only 5,730 years, which means that it is useful for determining the age of things between 100 and 70,000 years old. (In geologic time, 70,000 years is very, very recent.) This means that any sample older than 70,000 years will appear to be 70,000 years old — never older — either because there are no carbon-14 atoms left or there are not enough to measure with current techniques.

Relatively Absolute: Combining Methods for the Best Results

With the advancement of absolute dating techniques, you may think relative dating is no longer needed. However, combining relative and absolute methods of dating is useful for many reasons:

>> Absolute dating requires the presence of certain elements, which may not be found in all the layers of rock that scientists want to have dates for. In such cases, a scientist may need to rely on relative dating for portions of a rock sequence that can't be dated with absolute methods. The result is that geologists continue to rely on relative dating to fill in gaps for periods when they cannot measure accurate absolute ages.

>> Absolute dating of sedimentary rocks can be difficult because the sediments provide dates of the mineral's formation from molten rock, not the formation of the rock they are currently in. Therefore, a different approach must be used to date layers of sedimentary rock. This approach is called bracketing.

Bracketing means that the geologist takes a sample of dateable material (for example, a mineral from a metamorphic or igneous rock) from immediately above and immediately below the rock layer (or rock) she wants an age for. By measuring dates on these samples, she can determine a minimum age and maximum age for the rock layer in question. This approach is more specific than methods of relative dating but still does not give the exact age of the rock. Rather, it provides a bracket of "somewhere between this age and that age."

>> Methods of absolute dating require intensive work in the laboratory, many years of training, or money to pay a special scientist to do the analysis for you. It would be incredibly time-consuming and expensive to sample and date every single rock or rock layer on the earth's surface. Instead, geologists choose to sample and study specific parts of the rock record — layers that seem interesting or significant in some way, such as indicating the boundary between two time periods when something dramatic happened (like the extinction of the dinosaurs).

With the combined approach, scientists focus their time and money on specific, important questions that require absolute dates, and they still depend on relative dating for other parts of the rock record.

Eons, Eras, and Epochs (Oh My!): Structuring the Geologic Timescale

Long before they developed methods of absolute dating, geologists had composed a geologic timescale based on relative dating and stratigraphy. The timescale was first organized into segments based on changes in which fossils were present in each layer and other relative methods of dating. When techniques of absolute dating were discovered, geologists applied methods of absolute dating to provide ages for each segment on the timescale.

The longest time spans on the timescale are called *eons.* Beginning with the Hadean or Pre-Archean eon, the different eons mark dramatic changes on Earth, such as the appearance of life (the Archean), the formation of an oxygen-rich atmosphere (the Proterozoic), and later the spread of complex life (the Phanerozoic, the current eon). Sometimes you will find the Hadean, Archean, and Proterozoic eons collectively called the *Pre-Cambrian.*

Each eon is split into multiple *eras.* The eras often indicate a change in the dominant animals in the fossil record. For example, the Paleozoic was ruled by marine animals without spines *(invertebrates),* the Mesozoic was the era of the dinosaurs, and the Cenozoic (the current era) is the age of mammals.

Each era is then split into multiple *periods,* and the periods are separated into *epochs.* Each division indicates an important change in the living organisms represented by fossils in the rocks. That's because long before geologists had methods of absolute dating, using the fossils was the best way to segment Earth's history. Due to the nature of the geologic record, geologists have collected much more detail about events and fossils that occur closer to recent time. This becomes evident in the organization of the geologic timescale, where the closer you get to modern time, the smaller the divisions are and the more details the periods and epochs have.

Figure 16-7 illustrates the geologic timescale with the segments most commonly used by North American geologists and absolute dates for the boundaries between them. Keep in mind that each era has many periods, and each period has multiple epochs, even though they are not all illustrated here.

The result of developing a relative timescale before an absolute timescale is that the segments of the geologic timescale are not separated into equal periods of time. At each level, the separations occur as the fossils indicate biological change, not along any set number of years. For example, the Archean eon spans nearly 1.5 billion years, while the Phanerozoic spans 542 million years. This variation is true even in the more recent epochs; for example, the Miocene spans about 18.4 million years while the following Pliocene epoch spans only 3.7 million years.

While the inconsistent number of years in each segment makes it challenging to learn the details of the geologic timescale, the division between each segment represents a meaningful change in the geologic and fossil records. This approach ends up being much more informative than evenly spaced intervals of time (like 100 years or 1,000 years) that have no meaning.

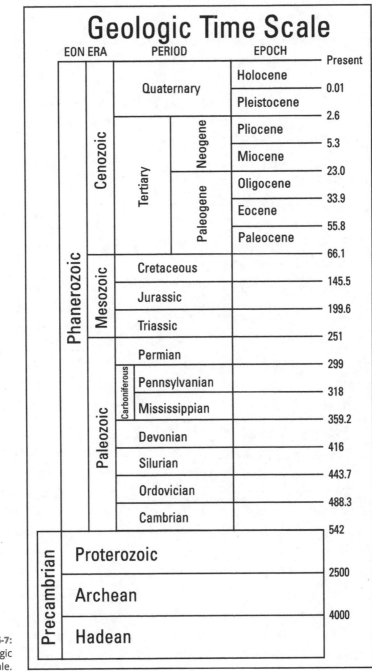

Geologic Time Scale

EON	ERA	PERIOD		EPOCH	
					Present
		Quaternary		Holocene	
					0.01
				Pleistocene	
					2.6
	Cenozoic	Tertiary	Neogene	Pliocene	
					5.3
				Miocene	
					23.0
			Paleogene	Oligocene	
					33.9
				Eocene	
					55.8
Phanerozoic				Paleocene	
					66.1
	Mesozoic	Cretaceous			
					145.5
		Jurassic			
					199.6
		Triassic			
					251
	Paleozoic	Permian			
					299
		Carboniferous	Pennsylvanian		
					318
			Mississippian		
					359.2
		Devonian			
					416
		Silurian			
					443.7
		Ordovician			
					488.3
		Cambrian			
					542
Precambrian	Proterozoic				
					2500
	Archean				
					4000
	Hadean				

FIGURE 16-7:
The geologic
timescale.

IN THIS CHAPTER

» **Understanding what evolution explains**

» **Following the development of the theory of evolution**

» **Using the fossil record to challenge predictions**

» **Preserving body parts and behaviors**

» **Considering bias in the fossil record**

Chapter **17**

A Record of Life in the Rocks

t may seem strange to find a chapter about biological change and evolution in the middle of your geology book. But it makes perfect sense if you consider that the record of life in the rocks provides answers to questions about evolution and long-term biological change. When scientists look at the geologic record, they find evidence and support for biological theories of change, such as evolution and the relationships among different organisms.

In this chapter, I briefly explain the current theory of evolution, its development over the last 150 years, how the geologic record provides a laboratory for testing this theory, and how the preservation of past organisms in rocks allows scientists to reconstruct a detailed history of evolution — of biological change through time.

Explaining Change, Not Origins: The Theory of Evolution

In Chapter 2, I explain exactly what scientists mean when they call something a theory. To summarize, a scientific theory is not simply an assumption or a best guess. A *scientific theory* has been tested repeatedly and accepted as an accurate explanation of a complicated phenomenon, such as the theory of plate tectonics I explain in Part 3. The *theory of evolution* explains how biological organisms change through time.

A common misconception about the theory of evolution is that it explains how life on Earth began. This is false. Scientists have not yet figured out how life on Earth began. What they have documented through observations and experiments, and with the help of the geologic record, is how living organisms transition from one species to another over long periods of time.

REMEMBER

The theory of evolution explains the mechanism of biological change, or *how* this change occurs, not why it happens. The mechanism for biological change in organisms is what I describe in this chapter.

The Evolution of a Theory

Like all scientific theories, including the theory of plate tectonics described in Chapter 9, the theory of evolution began with ideas that have since been proven wrong or expanded and improved upon. In this section, I briefly describe the development of the modern theory of evolution.

Acquiring traits doesn't do it

When scientists first began to explore ideas about the inheritance of physical characteristics, or *traits*, some suggested that an organism could change a physical trait during its lifetime and then pass this modified trait on to its offspring.

A naturalist named Jean-Baptiste Lamarck observed giraffes and proposed the following scenario to explain their long necks. He suggested that a giraffe that needs to reach leaves at the top of a tree will stretch its neck. After its neck is stretched, the offspring of this giraffe will be born with longer necks and, therefore, the ability to reach leaves on the tallest trees. He called this idea the *inheritance of acquired traits*. Lamarck was correct in his inference that traits are inherited, but he was incorrect about how the traits are acquired.

As an example, consider the modern science of laser eye surgery. Perhaps for many generations, everyone in your family has needed to wear glasses. You decide to correct this problem once and for all by getting corrective laser eye surgery. You no longer need glasses, but it is still very likely that your children will need glasses. Your corrective eye surgery, a trait acquired during your lifetime, does not transfer to your children.

Naturally, selecting for survival

In the first half of the nineteenth century, two naturalists — Alfred Wallace and Charles Darwin — were independently developing the same idea about biological change and didn't know it. Darwin eventually published his ideas in the book *On the Origin of Species* and has become known for the idea of natural selection, even though Wallace was also working on the same ideas at the same time.

In Darwin's time, plant and animal breeders were already practicing artificial selection, much like breeders do today. In *artificial selection*, the breeder chooses an organism with traits that are desirable and carefully breeds it to produce a specific result. Examples of desirable traits would be beautiful color in flowers and particular flavors in fruit.

Understanding how artificial selection resulted in offspring with particular traits, Darwin proposed that a natural process of selection worked similarly in nature. He proposed that some traits may be more useful than others to an organism's survival and that these traits would be naturally selected for, leading these traits to be passed on to the next generation.

To use the example of a giraffe again, Darwin suggested that a giraffe with a longer neck can reach leaves on higher branches than giraffes with shorter necks. This long-necked giraffe therefore has access to more food. The success of the long-necked giraffe at finding food increases its chance of survival and its opportunities to mate, reproducing offspring with similarly long necks.

Darwin and Wallace did not explain exactly how traits are passed from a parent to its offspring, but other scientists at the time were asking that very question.

Mendel's peas please

In the mid-1860s, a monk named Gregor Mendel experimented with pea plants and determined that characteristics (such as flower color) seem to be controlled by a pair of factors, one from each parent. Mendel determined that if the two factors in the pair are different, one may be expressed in the offspring while the other remains present in the cells and may be expressed in future offspring. These days, scientists know that the "factors" Mendel proposed are actually genes (which I describe next).

For example, consider eye color. Maybe you have a grandparent with blue eyes but both your parents have brown eyes. It is still possible for you (or your children) to have blue eyes: an expression of the blue-eyed gene that is present, though perhaps not expressed, in every generation.

Genetic nuts and bolts

In the early twentieth century, scientists determined that within their cells, all organisms have molecules of DNA (deoxyribonucleic acid) organized as *chromosomes*. Each cell in an organism's body has two sets of the chromosomes needed to create that organism, with one exception: the sex cells. The sex cells, sperm and egg cells, each have only one set of chromosomes.

REMEMBER

When an organism reproduces sexually (by combining sex cells from two separate individuals), the single set in the egg cell and the single set in the sperm cell are combined to form a new, unique pair of chromosomes, resulting in a new (and unique) combination of characteristics in the offspring.

Spontaneously mutating genes

The recombination and passing down of chromosomes from parent to offspring explains how current traits are inherited, but it does not explain how new traits appear. Something must change within the DNA to create a new trait, and how is this change possible? It's possible through genetic mutation.

REMEMBER

Genetic mutation occurs when the DNA of a chromosome changes in some way. A mutation may be caused by a *mutagen* — a chemical that alters the chromosome — or it may occur randomly and spontaneously. Whether the mutation is good, bad, or neutral depends on how it is expressed as a trait or physical characteristic and whether that trait is then useful for an organism's survival:

>> **The good:** If the mutation expresses as a trait that provides the organism a way to better adapt to its environment, or a way to compete strongly with other organisms for resources (such as a longer neck among giraffes), it is a *beneficial mutation.* A beneficial mutation improves the organism's chance of survival and opportunity to reproduce, and therefore the mutation is passed down to the next generation as part of the genetic code in the chromosomes.

>> **The bad:** If, however, the mutation produces a trait that prevents the organism from competing successfully or gets in the way of its survival, the chance of that organism surviving to reproduce is unlikely. For example, mutations that lead to disease or physical defects may shorten the life of an organism, not providing it a chance to reproduce. Therefore, the mutation will not be passed on to future generations.

>> **The neutral:** Some mutations are neither good nor bad; they simply occur, having no direct effect — positive or negative — on the organism's survival. This can be because the mutation does not express as a physical trait, or because the trait is neither helpful nor damaging for the organism's survival. These are called *neutral mutations.* For example, some domestic cats have six toes, which neither helps nor hinders their survival.

Speciating right and left

As chromosomes accumulate random mutations and physical traits change in response, the organism survives and reproduces according to how beneficial those traits turn out to be. Over a long period of time, this accumulation of beneficial traits in a population, or group of reproducing individuals, may change the physical characteristics of that population so much that a new species is created. The development of new species from existing ones through mutation and natural selection is called *speciation* and is the basis of evolution via natural selection, or change in organisms, through time.

It may be helpful to remember the following key points about how evolution through natural selection works:

>> Genes mutate randomly.

>> Physical traits are expressed.

>> Individuals are selected.

>> Populations reproduce and evolve.

TIP

A *species* is defined in biology as a population of individuals that breed in nature and produce fertile offspring. This definition rules out animals such as the *liger*, a combination tiger and lion that is bred in captivity but is sterile. Ligers do not compose a new species.

Two competing explanations exist for how quickly new species develop through the process of natural selection:

>> **Punctuated equilibrium:** The *punctuated equilibrium* explanation proposes that very little change takes place until a sudden surge of changes occur, creating new species over a relatively short time, such as a few thousand years.

>> **Phyletic gradualism:** *Phyletic gradualism* suggests that a gradual accumulation of minor genetic changes occurs continually through time, slowly creating new species.

Modern understanding of evolution is informed by scientists from many different fields. Together, through testing and experimentation, they build strong support for the theory of evolution as a mechanism for biological change.

REMEMBER

The combination of information from scientific specialists including geneticists, biologists, and paleontologists (who study fossil remains of organisms) is called *the modern synthesis* or *Neo-Darwinian evolution*.

Modern biological research in particular has shown that although life on Earth is very diverse, all living organisms share certain basic characteristics. For example, all life is composed primarily of four elements: carbon, nitrogen, hydrogen, and oxygen. And all living things have DNA in chromosomes that are passed on during reproduction.

These similarities suggest that the wide variety of living things currently existing on Earth share a very, very distant common ancestor.

Putting Evolution to the Test

The geologic record — specifically *fossils*, the preserved remains of once-living organisms — provides an active laboratory to ask and answer questions about evolution. Based on the theory of evolution, scientists can make predictions about what they expect to find in the fossil record, and then they test their predictions by gathering evidence from the rocks.

REMEMBER

Here are examples of some predictions based on the theory of evolution that can be tested with the fossil record:

>> If evolution has taken place, the oldest rocks should have remains of organisms very different from organisms today.

>> If organisms today are descended from organisms in the past, there should be intermediate fossil forms of organisms that link the two.

>> If evolution occurred, organisms that appear related today should have a common ancestor in the fossil record and show increasing differentiation from that ancestor through time to the present.

Keep reading through the rest of Part 5 to find out how each of these predictions has been proved true by fossil evidence in the geologic record.

Against All Odds: The Fossilization of Lifeforms

In order to answer questions about past organisms and their evolutionary relationship to modern plants and animals, scientists examine fossils preserved in the geologic record. *Fossils* may be actual remains, such as bones, or traces of past behaviors, such as footprints or tracks.

Bones, teeth, and shell: Body fossils

Body fossils are preserved body parts. Body fossils can be preserved in different ways. In some cases the body parts remain unchanged and are simply preserved as they existed. These are called *unaltered body fossils.*

Unaltered body fossils are rare and may result from freezing or mummification. In both cases the skin and soft tissue, as well as the hard body parts (such as bone), can be preserved. Another method of unaltered body fossil preservation is entrapment in a thick fluid, such as tar or tree sap (which hardens into amber).

More commonly, fossils in the geologic record are *altered* body fossils. In this case, the remains may be changed through a chemical process, preserving the shape of the body part but not its original composition.

Altered body fossils are preserved in three basic ways:

>> **Replacement:** *Replacement* fossilization occurs when remains of an organism (usually the hard parts, such as shell or bone) are buried in sediments, dissolved, and replaced by new minerals.

>> **Permineralization:** Preservation by *permineralization* occurs when minerals seep into the open spaces of buried remains, such as bones or wood, but leave some of the original organic material in place.

>> **Carbonization:** Some organic materials, such as leaves or insects, can be preserved as *carbon film.* This preservation results when all that remains of the original organism is a thin film of carbon preserving its shape.

REMEMBER

By far, the most common type of fossil preservation is in the form of a mold or a cast. A *mold* forms when an organism is buried in sediments, the sediments are hardened into a rock, and the remains of the organism decay or dissolve away. What remains in the rock is a mold of the organism but no actual parts of the organism. A *cast* forms when sediments fill the cavity of a shell or bone and harden, preserving the inner details of the body part.

Just passing through: Trace fossils

Some fossils provide evidence of an animal's activity in the distant past without preserving any part of the organism itself. These kinds of fossils are called *ichnofossils* or *trace fossils* because the organism has left only a trace, or small indicator, of its life and behavior. Trace fossils are any preserved indication of an organism's activity and include the following:

>> **Burrows:** *Burrows* indicate how an organism lived. Small burrows of ocean-dwelling organisms, for example, are easily preserved in the soft sediments at the bottom of the sea.

>> **Tracks and trails:** Organisms that move across land or along the ocean bottom may leave *footprints, tracks,* and *trails* in the sediments. They may be buried and become trace fossils, indicating movement and motion.

>> **Coprolites:** *Coprolites* are fossilized droppings, or feces, from an organism.

Figure 17-1 illustrates burrow and track trace fossils.

REMEMBER

The challenge with trace fossils is that scientists cannot know for certain which animal created the clues that are left behind (unless that animal is preserved in the act of creating the trace fossil!). But trace fossils are still important because unlike body fossils of organisms, trace fossils provide clues to the past actions, habits, and living patterns of ancient organisms.

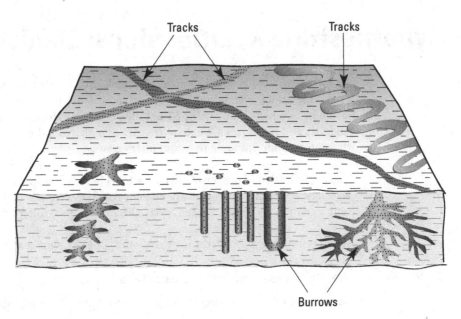

Tracks Tracks

Burrows

FIGURE 17-1:
Trace fossils include tracks and burrows.

Correcting for Bias in the Fossil Record

While fossil remains of organisms are useful to explore hypotheses about early life, the evolution of life, and other questions about Earth's biological past, you should always keep in mind that the fossil record is incomplete, or biased.

By *biased* I mean that only part of the story is being told while other parts are being left out or overlooked. In the case of the fossil record, the story told in the rocks is almost entirely about organisms with bones, teeth, or shells. These hard body parts are more often and more easily preserved. This means that some organisms (such as dinosaurs and shellfish) are overrepresented, while others (such as jellyfish or earthworms) are underrepresented or absent.

REMEMBER

The fossils that are preserved represent only a very small portion of all the creatures that have existed on Earth. The history of life in the rocks is biased toward those creatures who were in the right place at the right time and had the right features.

Add to that fact the many dramatic geologic events that shift, crack, stretch, and uplift the rocks on Earth's surface, exposing existing fossils to be weathered and disappear forever. You can see why the fossil record illustrates only a small fraction of the extraordinary variety of life that has inhabited the planet in the last 4 billion years. Scientists recognize this challenge and keep it in mind when using the fossil record to answer questions about the long history of life on Earth.

Hypothesizing Relationships: Cladistics

Scientists who study fossils, called *paleontologists,* have developed a method for predicting (and then testing) hypotheses about evolutionary relationships among different organisms. This method is called *cladistics.* Through cladistics, paleontologists and biologists are able to classify organisms in the present and the past through an understanding of evolutionary change.

The classification of organisms in cladistics is based on *shared, derived traits:* characteristics of each animal that are present now but were not present in their distant, common ancestors. Shared, derived traits are also called *synapomorphies.*

When classifying animals through cladistics, scientists make a few assumptions:

>> Organisms are related by descent to a common ancestor.

>> Characteristics of a *lineage* (a line of related organisms) change through time.

>> Each time change occurs, the organisms are split into two groups: one with the old characteristic, and one with the new characteristic. (The split is called a *bifurcating pattern* in the lineage.)

It is important to note that while these are the assumptions made by cladistics, many scientists propose alternate or opposing hypotheses as well. For example, some scientists do not accept that lineages must split in two, displaying a bifurcating pattern. They suggest that multiple groups with different characteristics may arise at the same time, in which case cladistics would not be as useful in classifying evolutionary relationships.

REMEMBER

The power of cladistics is that it allows scientists to propose evolutionary relationships between species. These proposals are hypotheses that can then be tested and put through the rigors of the scientific method that I describe in Chapter 2.

The result of cladistic analysis is a chart illustrating the evolutionary relationship called a *cladogram* or *phylogenetic tree.* Two styles of phylogenetic trees are illustrated in Figure 17-2.

These cladograms illustrate a simplified version of the evolutionary relationship between some vertebrate animals (those with a backbone; see Chapter 19). Each split branch indicates a new characteristic that makes the descendant groups different from one another. For example, the first split indicates that ray-finned (or spiny) fish and lungfish shared a common ancestor. (I describe the evolution of fish in Chapter 19.) The descendants of this ancestor are *either* spiny *or* have lungs.

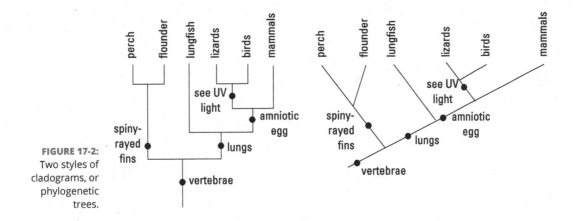

FIGURE 17-2:
Two styles of
cladograms, or
phylogenetic
trees.

Similarly, further up the tree, mammals and lungfish share an ancestor, but when the characteristic of mammals known as *amniote eggs* appears, the two groups branch apart. (I explain the importance of amniote eggs in Chapter 19.) The circles show you where on the branches these characteristics separate one type of animal from its closest relative. For example, mammals are more closely related to lungfish than they are to other fish. Another example is that lizards and birds are more closely related to each other than either of them is to mammals.

As you read the next chapters about Earth's history and the evolution of organisms through time, keep in mind that one of the reasons the fossil record is so important is that it allows scientists to test the hypotheses they've created through the use of cladistic analysis.

Chapter **18**

Time before Time Began: The Precambrian

The Precambrian time period of Earth's history is very mysterious. It occurred so long ago that most of the evidence for the processes that first took place on Earth have been destroyed. Almost no fossils and very few rocks are left from this period. The few rocks that do remain tell a very interesting story about Earth's beginnings as a planet.

In geologic time, the Precambrian covers the first 4 billion years of Earth's history. Geologists separate the time period into three eons: the Hadean, the Archean, and the Proterozoic. During this long expanse of time very important events took place, such as the creation of the first continental crust, the accumulation of water on the surface, the formation of an atmosphere, and the appearance of the earliest life forms.

In this chapter, I describe scientists' theories and hypotheses about how the earth (and solar system) formed, how the first continents were created, and what geologic processes (such as plate tectonics and the rock cycle) began billions of years ago and continue to occur today.

In the Beginning . . . Earth's Creation from a Nebulous Cloud

More than 4.5 billion years ago, the materials that constitute the sun, Earth, and other planets in the solar system were part of a big cloud of gaseous matter, called a *nebula*, swirling around in the Milky Way galaxy.

According to the *solar nebula hypothesis,* some of this matter collapsed into each other, becoming the giant ball of light and energy we call the sun. After the sun formed, the remaining matter was left as particles in a spinning disk of cloudy material around it. Small rocky particles in the cloud were attracted to one another by the force of gravity. As they crashed together, they became larger rocky objects, or *planetesimals,* flying in orbits around the sun.

Over time these planetesimals continued to attract matter or crash into each other, growing larger until they formed the rocky internal planets of our solar system: Mercury, Venus, Earth, and Mars.

Many scientists believe that shortly after the earth formed, another planetesimal, possibly as large as Mars, crashed into it, knocking a large amount of surface material off the earth and into orbit around it. This material eventually combined together to form the moon, which is held close to the earth by the pull of gravity. This is called the *Giant Impact Hypothesis* and continues to be tested today as scientists seek to refine their understanding of the relationship between the earth and the moon.

The newborn Earth was very different from the earth you are familiar with. Almost all the matter or elements that make up the modern Earth were present as atoms and molecules, but they were all mixed up or *undifferentiated.*

Under the influence of gravity's pull, the densest materials (such as iron and nickel) sank into the center of Earth's swirling ball of matter and formed the core. Surrounding the core, somewhat less dense elements collected, forming the mantle. The least dense materials were left to form the outermost layer, the lithosphere. These layers are described in detail in Chapter 4.

REMEMBER

This process of separation based on density and other chemical characteristics is called *differentiation,* and it created the distinct layers of varying composition observed in the modern Earth. In fact, differentiation continues today, moving less dense materials to the surface of the earth through the subduction, partial melting, and eruption of rock materials (as I describe in Chapter 9).

Scientists believe that much of the surface water on Earth today arrived with icy comets that crashed into the young Earth. Upon impact, the ice melted into water and remained trapped on Earth due to the pull of gravity. Another source of water on early Earth may have been water vapor erupting from volcanoes.

Addressing Archean Rocks

Soon after Earth's formation and differentiation, the lithosphere of the earth was separated into plates that began moving around the surface (similar to what scientists observe today). The oldest continental crust rocks are from the Archean eon, approximately 4 billion years ago. They provide a picture of the geologic processes that were occurring at that time. In this section, I describe some of the Archean rock formations, most of which were created about 2.5 billion years ago, and I explain how they provide evidence for plate tectonics so early in Earth's history.

Creating continents

As the heat generated by the formation of the earth and the radioactive decay of elements in its core escaped through volcanoes on the seafloor, rock materials that constitute the cores of our modern continents began to form. (For details on radioactive decay, see Chapter 16.)

The oldest rocks on a continent constitute its core, or *craton.* A craton is made of a *Precambrian shield* of ancient rocks that are visible on the surface, or exposed, and a surrounding *platform* of ancient rocks that have been covered with more recent rocks. Figure 18-1 illustrates the location of each continent's craton.

The craton of a continent is created when multiple volcanic arcs (which I describe in Chapter 9) crash together to form larger and larger land masses. Evidence for these tectonic processes is found in the Archean rocks of the continental cratons.

Revving up the rock cycle

The craton of each modern continent is composed of parallel formations of granite-gneiss complexes and greenstone belts. Together these rocks illustrate that all the processes observed in the rock cycle today were actively taking place by 4 billion years ago. Keep reading to find out how scientists have come to such a conclusion.

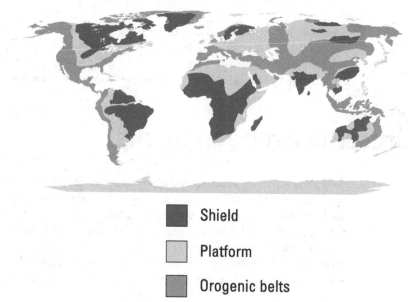

Shield

Platform

Orogenic belts

FIGURE 18-1: Cratons of the modern continents.

Granite- gneiss complexes

Geologists have observed on present-day Earth how rocks are formed and transformed by the processes associated with plate tectonics. (I describe these processes in Chapters 7 and 10.) The different plate boundaries and processes result in different rock types and characteristics. For example, mantle upwelling at mid-ocean ridges produces dense igneous rocks such as basalt, and continental uplift and the weathering of existing rock result in sediment particles being transported and deposited to form sedimentary rocks.

In a few of the continental cratons, geologists have found *granite-gneiss complexes* that they have dated to nearly 4 billion years ago. In Chapter 7, I note that granite is an intrusive igneous rock and gneiss is a metamorphic rock. For granite to form, there must be conditions of heat and/or pressure that melt rock into magma (I describe these processes in Chapter 10). These conditions are most commonly found at subduction zone boundaries between two converging plates. Gneiss is formed by the intense compression and deformation of preexisting rocks, such as might occur with subduction and continental collision Thus, the presence of granite-gneiss complexes is evidence for active tectonic processes, specifically convergent boundary processes as early as 4 billion years ago. (See Chapter 10 for more on tectonic processes and subduction.)

Greenstone belts

Another Archean rock formation, called *greenstone belts,* provides geologists with evidence of crustal rocks being weathered and forming sedimentary rocks by 2.5 billion years ago (a process I describe in Chapter 7). Greenstone belts are metamorphosed mafic igneous rocks similar to ocean ridge basalt rocks today. (The name of these rocks comes from the green color of the mineral chlorite that forms as basalt experiences metamorphic conditions.) Associated with these metamorphic rocks are layers of sedimentary rock. Sedimentary rocks can form only when a preexisting rock is exposed to the atmosphere and is weathered into sediment particles. The presence of sedimentary rocks with metamorphic rocks in the greenstone formations indicates that processes of uplift (causing metamorphism) and erosion (producing sediments) were active in the Precambrian time period.

Piecing observations together

Geologists apply the theory of uniformitarianism (which I describe in Chapter 3) and hypothesize that granite-gneiss complexes and greenstone belts are the result of crustal plates moving apart from each other and then moving back together repeatedly. Most greenstone belts are located in linear, parallel arrangements, separated by granite-gneiss complexes. One hypothesis suggests that the parallel arrangement is the result of activity along a plate boundary that experienced cycles of rifting, convergence, subduction, and volcanism. The associated granite-gneiss complex would be a result of intrusive igneous rocks (granite) and rocks transformed by compression and heat (gneiss) as the plates moved together whereas the greenstone indicates the uplift and erosion parts of the rock cycle were occurring.

Multiple repetitions of this sequence of events may seem impossible, but over a period of *billions* of years, they are entirely possible.

Feeling hot, hot, hot: Evidence for extreme temperatures

Scientists have found evidence in the ancient rocks suggesting that the internal temperature of the earth was much hotter than today and has been slowly cooling off over the last 4 billion years. When the earth first formed, the coalescing of all that matter generated a large amount of heat that became trapped inside the earth. In the process of cooling, the heat from inside the earth escaped to the surface and radiated out, away from the planet into space. This process still occurs today, but the levels of heat are much lower than they were billions of years ago. Much of the heat generated by Earth today is believed to be the result of radioactive decay of elements in Earth's core.

In the Archean greenstone formations, geologists find really dark, dense, iron-rich volcanic rocks called *komatiites*. Komatiites have high amounts of iron in them, indicating that the magma (molten rock) from which they were formed must have been hot enough to melt iron-rich minerals. (In Chapter 7, I explain how different minerals melt at different temperatures.) This means the magma was much hotter than the magmas that erupt as lava on the surface of the earth today. Komatiites must have cooled from molten rock that was at least 1,600° C when it erupted onto the surface of the earth. For comparison, consider that today the highest temperature recorded for a surface flow of lava is 1,350° C.

With temperatures this high just below the surface of the earth (where magma forms), the mantle of the earth must have been even hotter. Scientists conclude that higher mantle temperatures might have led to faster mantle convection (see Chapter 10) and more active tectonic processes at plate boundaries. Except for the greenstone and granite-gneiss formations, very little evidence of these early, extremely active tectonics remains due to the continued subduction and recycling of Earth's crustal plates over the millions of years since the Precambrian ended.

Originating with Orogens: Supercontinents of the Proterozoic Eon

During the Proterozoic eon (between 2.5 billion and 542 million years ago), the young continents moved around the earth crashing into each other. Landmasses grew, rocks were deformed, and elevations changed through a sequence of *orogenies*, or mountain-building episodes.

REMEMBER

The long, linear rock formations that are deformed by each continental collision are called *orogens*. By mapping orogens, geologists have constructed a history of some of the events that created modern continents.

Over a period of almost 2 billion years, plate movements continued, building early continents such as *Laurentia,* which contained pieces of modern Greenland and Scotland attached to the craton of North America. Laurentia was formed by the collisions of plates *(orogenies)* and by *accretion,* or the addition of rock material from the crustal plates that it collided with.

Laurentia was a large landmass, but it was not a supercontinent. A *supercontinent* is a single landmass consisting of at least two preexisting landmasses. While Laurentia contained pieces of more than two modern continents, it was only one of the Precambrian continents. (Numerous smaller land masses, including Baltica, Siberia, and others were also present.)

By approximately 1 billion years ago, all the Precambrian continents had crashed together and formed the first supercontinent, *Rodinia.* The evidence for the formation of Rodinia is found in rocks of a substantial mountain-building event called the *Grenville Orogeny* dated to between 1.3 and 1 billion years ago. The rocks of the Grenville Orogeny are found on the cratons of all the modern continents (illustrated in Figure 18-1). The Grenville Orogeny is responsible for accreting large portions of continental material onto Laurentia as it formed Rodinia. In fact, when Laurentia was part of Rodinia, 75 percent of the materials that form the modern North American continent were already present.

REMEMBER

It is possible that supercontinents existed prior to Rodinia, but no evidence of earlier supercontinents exists in the geologic record.

Between 750 and 650 million years ago, the landmasses of Rodinia broke apart and reconnected in a different arrangement known as the supercontinent *Pannotia*. Pannotia lasted for only about 60 million years before breaking into the four major landmasses that created the largest supercontinent, Pangaea, at the end of the Paleozoic era. (I offer details on Pangaea in Chapters 19 and 20.)

Single Cells, Algal Mats, and the Early Atmosphere

As you know, without green plants (including algae) you would have no oxygen to breathe. The modern Earth is covered with plant life that transforms carbon dioxide into oxygen so that organisms, including humans, can exist. This wasn't always the case. In this section, I explain how scientists think the first atmosphere developed and what role the earliest life forms may have played in creating the type of atmosphere you enjoy today.

Hunting early prokaryotes and eukaryotes

Scientists don't know how the first life on Earth originated. What they do know is that there is evidence for living organisms in rocks from 3.5 billion years ago. This earliest life was *prokaryotic* or simple-celled. Prokaryotes were (and still are) small cells with no internal organs that reproduced asexually (without the recombination of genes, as I explain in Chapter 17). Modern-day prokaryotes include bacteria.

Evidence also exists that slightly more complex cells called *eukaryotes* emerged about 2 billion years later (about 1.5 billion years ago). The earliest eukaryotes were also single-celled, but their single cell contained structures including a cell nucleus, and scientists believe that they reproduced sexually, like modern eukaryotes (such as humans, animals, and plants) do. A comparison of the cell structure between prokaryotes and eukaryotes is illustrated in Figure 18-2.

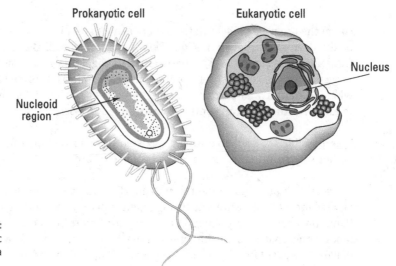

Prokaryotic cell

Eukaryotic cell

Nucleus

Nucleoid region

FIGURE 18-2:
A prokaryotic cell and a eukaryotic cell.

You know it as pond scum: Cyanobacteria

A prokaryotic organism that you may be familiar with is *cyanobacteria*. Cyanobacteria is a single-celled organism that also goes by the name *blue-green algae*. When large colonies of cyanobacteria flourish, they create a slimy green film on the surface of a lake or pond; thus, they have also become known as *pond scum*.

REMEMBER

These simple, single-celled organisms are extremely powerful. They possess the ability to transform sunlight and carbon dioxide gas into sugar energy and oxygen gas via a process called *photosynthesis*. As Figure 18-3 illustrates, photosynthesis is the biological process through which green plants transform sunlight, water, and carbon dioxide into oxygen and energy. The plants use the energy to grow and release the oxygen back into the atmosphere.

Scientists think that by the middle to the end of the Archean eon, communities of cyanobacteria were practicing photosynthesis. Fossilized structures called *stromatolites* provide evidence that these photosynthetic organisms existed by the beginning of the Proterozoic eon.

REMEMBER

Stromatolites are mounded structures built by colonies of algae as they grow upward toward the sun. Stromatolites form when tiny strands of algae trap particles of sand and other sediments between them (see Figure 18-4A). As the algae die, they form layers of sediment from the trapped particles (see Figure 18-4B). New algae grows on top of them, up toward the sunlight (see Figure 18-4C). Over time a mound of sediment and algae layers forms (see Figure 18-4D). You can see a fossil stromatolite in this book's color photo section.

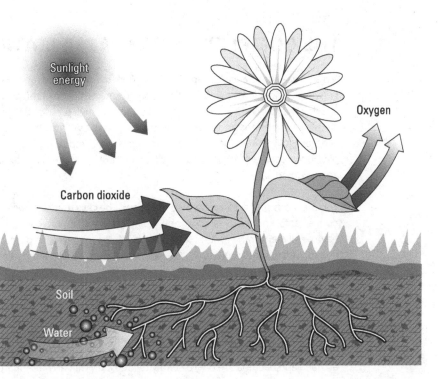

FIGURE 18-3:
The biological
process of
photosynthesis.

Sunlight energy

Oxygen

Carbon dioxide

Soil

Water

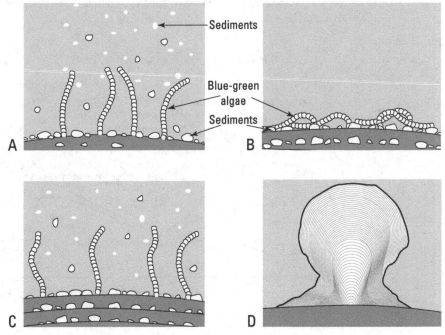

Sediments

Blue-green algae

Sediments

A

B

C

D

FIGURE 18-4:
Formation of a
stromatolite as
algae strands
trap sediments.

THE COMPLEXITY OF CLASSIFICATION

Modern genetic science has opened new doors to understanding the relationships between ancient life and modern living organisms. For a long time biologists organized all living things into six kingdoms: animals, plants, fungi, protists, bacteria, and archaea. Of these six, bacteria and archaea are the only ones that include simple-celled organisms, or prokaryotes. The other four kingdoms are composed of organisms with complex, eukaryotic cells. (Protists are *single*-celled, but their one cell is complex; thus they are eukaryotes.)

You are a complex and multi-celled organism. The cells that compose your body, your houseplants, and the mushrooms on your pizza are all eukaryotic cells. Eukaryotic cells have a cell membrane around them and multiple organelles inside, including a nucleus and mitochondria. In contrast, bacteria and archaea are single-celled organisms with prokaryotic cells. Prokaryotic cells are much smaller than eukaryotic cells and do not have a thick cell membrane or any organelles.

Based on this similarity alone, it seems logical that archaea and bacteria are more closely related to one another than they are to any plants, animals, or other eukaryotes. Modern genetic analysis of the ribonucleic acid (RNA), however, indicates that archaea and bacteria are only distantly related, and that archaea share a more recent common ancestor with eukaryotes. This relationship is illustrated on the branching tree diagram that includes the major families within the three domains: Bacteria, Archaea, and Eukarya.

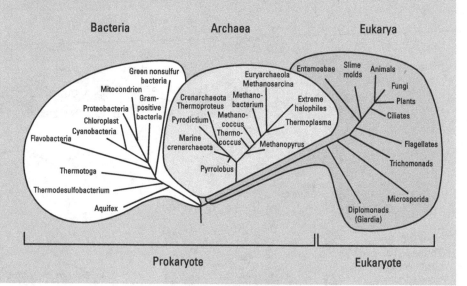

This finding illustrates a truth about biological classification that scientists have slowly been discovering: Common physical characteristics are not the strongest line of evidence for drawing conclusions about evolutionary relationships. Currently, scientists accept the classification of all living things into the three domains of Bacteria, Archaea, and Eukarya instead of the six kingdoms.

Stromatolites still exist today but are rare. One place they can be found is in warm, shallow, extremely salty ocean water, such as in Shark Bay in Western Australia, which is pictured in the color photo section. This environment is inhospitable to other creatures that might eat the algae, so the algae flourish without any predators.

Scientists think that the cyanobacteria that built Archean and Proterozoic stromatolites are largely responsible for the photosynthesis that added oxygen to the atmosphere around 2.3 billion years ago.

Waiting to inhale: The formation of Earth's atmosphere

The earth's early atmosphere originated from volcanic *outgassing*, or eruptions of gas and water vapor. The gasses that filled the atmosphere included water (H_2O) and carbon dioxide (CO_2) with very little "free" oxygen (O_2). There are two ways to separate oxygen from water and carbon dioxide molecules: photosynthesis and photochemical dissociation. In the previous section, I cover photosynthesis (courtesy of cyanobacteria).

Photochemical dissociation occurs when rays of sunlight hit water molecules in the atmosphere and break the molecular bonds. Once broken, the free oxygen atoms bond with other oxygen atoms to create ozone (O_3), as well as free oxygen molecules (O_2). The layer of ozone in the atmosphere protects the earth from the intense heat of the sun's radiation.

REMEMBER

While photochemical dissociation produces some free oxygen, after the ozone layer builds up, it prevents further photochemical dissociation. This means that photosynthesis must have played a larger role in increasing oxygen levels in the atmosphere toward the end of the Precambrian time period.

You may be wondering how it is possible to know what gasses were in the atmosphere billions of years. The answer is in the rocks!

A group of Precambrian sedimentary rocks called *banded iron formations* (BIFs) exist that provide clues to the evolution of Earth's atmosphere. BIFs have alternating layers of iron minerals and *chert* (a chemical sedimentary rock made of silica; see Chapter 7). Most likely, these rocks were created by the settling of particles and precipitation of minerals out of seawater.

BIFs present a puzzle to scientists studying the early Earth because for these rocks to form, iron must have been dissolved in seawater, which can happen only when almost no oxygen is present in the water. After the iron is dissolved into the water, however, the oxygen levels increase. When dissolved iron comes into contact with oxygen in the water, it forms the minerals hematite (Fe_2O_3) and magnetite (Fe_3O_4) found in the BIF layers.

GOING TO EXTREMES

For many decades, scientists seeking to understand the origins of life thought that only a limited range of conditions and environments were suitable to support life. They defined the range according to the needs of life as they knew it, which they understood to be carbon-based life. Carbon-based life requires access to water, the presence of carbon and oxygen, and a fairly narrow range of temperature and acidity. Imagine scientists' astonishment when they discovered life surviving in all the most extreme environments on the planet!

Extreme environments include places that are really hot, like the hot springs at Yellowstone National Park, or really cold, like the Antarctic ice sheet. Other environments considered extreme include hot water *(hydrothermal)* vents deep in the ocean that spew water rich in dissolved minerals and lakes super-saturated with salts, such as Mono Lake in California.

In recent decades, scientists have discovered that life does indeed exist in places previously thought too extreme to support life. Various species of bacteria and archaea thrive in these environments. As a group, these bacteria are called *extremophiles*. Not only can extremophiles exist in these environments, but also each species has evolved to be specifically adapted for its extreme environment.

Perhaps the most fascinating thing about the discovery of extremophiles on Earth is how it redefined the conditions needed to support life. This means that scientists looking for signs of life outside of the earth, or *extraterrestrial* life, can expand their search to include environments they previously assumed could not support any type of life.

Some scientists conclude that the formation of BIFs indicates that atmospheric oxygen levels at the end of the Archean eon and into the Proterozoic eon were fluctuating — increasing and decreasing and increasing again. When oxygen levels in the atmosphere (and therefore also the oceans) are low, iron minerals dissolve into the seawater. As the oxygen levels rise (most likely due to increased photosynthetic activity of algae), the iron combines with oxygen to form minerals that precipitate to the seafloor, depositing layers that eventually become rocks.

Another hypothesis that is being explored is that hydrothermal vents on the seafloor called *black smokers* may have contributed dissolved iron to the oceans that later deposited to form BIFs. Modern observation of black smokers shows that the heated water coming from them carries dissolved iron minerals that deposit, forming chimney-like shapes on the seafloor. Similar vents may have contributed to the amount of dissolved iron in ancient oceans, and along with changes in atmospheric oxygen, contributed to the cycle of BIF deposition. Today, banded iron formations are found in the Pilbara region of Australia and the Mesabi mountain range in Minnesota.

REMEMBER

The conditions to form BIFs are limited to between 2 and 2.5 billion years ago. After that time, oxygen levels in the atmosphere appear to have been steadily increasing. Evidence for this increase exists in rock formations called continental red beds. *Continental red beds* are sedimentary rocks of sandstone or shale, colored red due to the iron minerals such as hematite (Fe_2O_3).

As I explain in Chapter 7, sedimentary rocks such as sandstone or shale are the result of rock particles on land being weathered and transported to the sea. Unlike the BIFs, which are sedimentary rocks precipitated from seawater, continental red beds indicate that sediment particles on land were exposed to high enough levels of oxygen in the atmosphere to form hematite minerals.

Continental red beds appear after BIFs become rare, around 1.8 billion years ago, and are commonly found in Phanerozoic rocks created in the 542 million years since the end of the Precambrian. (To find out about what the earth was doing during the Phanerozoic eon, be sure to read Chapter 19.) Table 18-1 organizes the important things to remember about BIFs and continental red beds.

TABLE 18-1 **Comparing Banded Iron Formations (BIFs) and Continental Red Beds (CRBs)**

Rock Name	Rock Type	Evidence for Oxygen	Age
BIFs	Chemical precipitate sedimentary	Fluctuating levels of oxygen in the oceans	2.0–2.5 billion years ago
CRBs	Sandstone/shale sedimentary	Increasing oxygen in the atmosphere	After 1.8 billion years

SNOWBALL EARTH

Can you imagine Earth so cold that the oceans freeze over and ice covers all the continents? The relatively recent snowball earth hypothesis proposed in the early 1990s suggests that this may have occurred more than once during the Proterozoic eon. The hypothesis states that the entire earth was covered at least once, possibly many more times, with ice and snow. There are multiple lines of geologic evidence suggesting extensive glaciations, but they are not solid enough for the scientific community to form a consensus concerning the reality of, or the extent of, snowball earth conditions.

According to the hypothesis, after the ice covered large portions of the earth, the effect of *albedo*, where ice reflects the sun's energy rather than absorbing it, would result in rapid cooling of the earth's atmosphere and increase the growth of glaciers and ice sheets. When more ice is present, the albedo reflects even more energy back to space, decreasing the temperature further and allowing even more ice to build up. So after the earth was covered in ice, it would stay that way for millions of years until greenhouse gasses emitted from volcanoes (thanks to plate tectonics) built up in the atmosphere, leading to global warming conditions that could melt the ice.

Currently scientists are working to test the snowball earth hypothesis and are coming up with many questions that need to be answered before this hypothesis can be accepted as an explanatory theory. One question that many researchers are asking is what happened to life during the snowball earth episode(s)? Some scientists suggest that hydrothermal vents deep in the ocean, or warm water surrounding active volcanoes, provided a refuge for life to exist while the rest of the planet was frozen. Other researchers go so far as to suggest that the snowball earth event(s) played an important role in driving the evolution of complex cells (eukaryotes), but they have yet to find any supporting evidence for this assertion in the fossil record. And still others suggest that the earth could never have been completely covered in ice, that regions near the equator likely remained ice-free, at least during part of the year.

Questioning the Earliest Complex Life: The Ediacaran Fauna

For many years, geologists defined the end of the Proterozoic eon and beginning of the Phanerozoic (which means visible life) by the appearance of a wide range of shell-bearing fossils that date to 542 million years ago. The appearance of these fossils is called the *Cambrian Explosion* to document how dramatically diverse and sudden these life forms appear. For many decades, scientists have struggled to explain how such complex life could simply appear 542 million years ago without any, simpler precursor in the fossil record. However, a mid-century geologic find in Australia has caused scientists to reconsider this.

In 1946, a series of fossils were discovered in the Ediacara Hills of Australia. At the time, the fossil imprints of soft-bodied organisms were thought to be the remains of extinct jellyfish or marine plants. More than 30 different animal types were identified and described from the fossils, composing the *Ediacaran fauna.* Many of them have odd body shapes that are not seen in later Paleozoic (or modern) animals. Since first being identified in Australia, fossils of the Ediacaran fauna have been found on every continent (except Antarctica), and scientists are beginning to accept that these creatures may have been the dominant complex life on Earth's seafloors long before the Cambrian period began.

REMEMBER

The preservation of past life in the geological record relies on good conditions for preservation, as well as on preservable parts, like bones or shell. Without hard parts to fossilize, the early complex life of the Ediacaran fauna was overlooked, and only when it was recognized did it help answer one of paleontology's greatest mysteries.

Chapter **19**

Teeming with Life: The Paleozoic Era

The Paleozoic era of the Phanerozoic eon is marked by the appearance of shell-bearing organisms. Beginning approximately 542 million years ago, the biology of living organisms on the planet changed dramatically. Although simple, single-celled life had existed for billions of years, at the beginning of the Paleozoic the geologic record explodes with multi-celled shell-building life of all different shapes and sizes.

So many important changes occurred during the 300 million years of the Paleozoic era that it is often split into the early Paleozoic (the Cambrian, Ordovician, and Silurian periods) and the late Paleozoic (the Devonian, Carboniferous, and Permian periods). In this relatively narrow expanse of time (geologically speaking), organisms evolved from soft-bodied, simple sea creatures to fish, amphibians, reptiles, and eventually mammals. Meanwhile, photosynthetic algae gave rise to the first land-dwelling plants and trees. All this happened as the continents continued to crash into each other and break apart as the crustal plates moved around the surface of the earth — culminating in the formation of the most super of supercontinents.

In this chapter, I describe some of the most significant events of the Paleozoic era and broadly summarize the evolutionary trends and global geologic changes that took place.

Exploding with Life: The Cambrian Period

The beginning of the Paleozoic is marked by the sudden appearance of a wide variety of animal forms in the geologic record. In fact, the fossils from this period exhibit all the animal body plans that exist even today, 540 million years later. (A *body plan* is how an organism's body parts and growth patterns are organized.) This sudden appearance of complex life in the geologic record is called the *Cambrian Explosion,* and it's the focus of this section.

REMEMBER

The Cambrian Explosion has long been defined by the abundance of creatures preserved in the fossil record at the beginning of the Cambrian period. But it is likely that this sudden appearance of life is a result of the incomplete nature of a history told in rocks. Rather than documenting the first appearance of complex life (which was already present, as represented by the Ediacaran fauna I describe in Chapter 18), the Cambrian Explosion documents an important new adaption for life: the building of shells.

Toughen up! Developing shells

In the Cambrian, creatures living in the sea began to grow shells or *exoskeletons.* This gave them a tremendous advantage, both during their lifetimes and for being preserved in the geological record. As you can see today, this way of life has lasted for millions of years. External hard parts provide the following benefits:

>> **Protection from the sun:** During the Paleozoic, immense shallow seas were the primary habitat for life on Earth. Soft-bodied creatures were exposed to the sun's harmful rays — the same rays you and I avoid with sunscreen and hats. Building an exoskeleton protects the soft tissues and internal organs of a creature from being damaged by the sun.

>> **Moisture retention:** Large, shallow water environments sometimes experience an occasional absence of water — similar to the beach at low tide. Animals that become stranded when the tide goes out will dry out and die unless they have a shell that retains enough moisture to help them survive until the tide comes back in.

>> **Muscular support:** Building an exoskeleton provides a framework for muscles to attach themselves to. This function is the same one your internal

skeleton provides. Because an exoskeleton provides structure for muscle attachments, it allows an organism to grow larger than it would without such a supporting structure.

>> **Protection from predators:** Possibly the most important advantage a shell provides is protection from other animals that may attack and devour a soft-bodied creature.

REMEMBER

While all of these are great reasons to build an exoskeleton, scientists are not certain which advantage started the evolutionary trend toward external hard parts. Evidence in the fossil record of creatures with damaged shells indicates that they were being hunted, attacked, and probably eaten by predators. For some scientists, this fact is enough to conclude that predation was the driving force for the evolution of exoskeletons at the beginning of the Paleozoic.

Ruling arthropods of the seafloor: Trilobites

The first *shelly fauna*, or animals with exoskeletons, were tiny creatures; their shells were only a few millimeters in size. But it didn't take long for other animals to follow the shelly trend. The most famous creature of the Paleozoic — possibly its mascot — is the *trilobite*.

REMEMBER

Trilobites are *arthropods*, which today include insects, spiders, and crustaceans such as lobsters. Species of trilobites filled every nook and cranny of the ocean throughout the Paleozoic, but they did not survive the end–Permian extinction. (I discuss this and other extinction events in Chapter 22.)

BURGESS SHALE

The Burgess Shale is a rock formation in British Columbia containing preserved remains of creatures from the Cambrian period about 540 million years ago. It was discovered in 1909, but its importance wasn't realized until the 1960s. More than 60,000 fossils were recovered, many of which were arthropods (trilobites and similar organisms), but also specimens of other animal groups from the Cambrian Explosion. The best part of the Burgess Shale is that the fine-grained sediments of the shale preserved detailed features as well as complete specimens of soft-bodied organisms that would not have been preserved under other conditions. Geologists think an underwater mudflow covered the seabed and all the creatures in a sudden event, providing a snapshot in time of an undersea community of organisms from 540 million years ago.

You can visit the Burgess Shale site, now considered a World Heritage site, in Yoho National Park, British Columbia.

Trilobites had an exoskeleton that is segmented into parts, similar in appearance to roly polys (pill bugs) but with a horseshoe-shaped head segment. They ranged in size from itty bitty (just a few millimeters) to more than 50 centimeters (almost 2 feet) in length, but most of them were around 5 to 10 centimeters (about 3 to 4 inches). Some were blind, others had compound eyes (like some insects today), and certain species could roll up just like a roly poly, presumably for protection. A collection of trilobites are pictured in this book's color photo section and illustrated in Figure 19-1.

While trilobites continued to cover the seafloor throughout the Paleozoic, they were most diverse during the Cambrian period and begin to be overshadowed in the fossil record by the development of other types of creatures later in the Paleozoic.

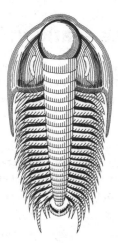

FIGURE 19-1:
A trilobite.

Building Reefs All Over the Place

Early Paleozoic life was lived under the sea. Without ice caps at Earth's poles, many of the large continents at the time were occasionally (meaning for a few million years at a time) covered by shallow seas, called *epeiric* or *epicontinental* seas. This expansive undersea environment meant that shallow marine and reef-building organisms would dominate the earth for millions of years.

The extensive shallow sea environments of the early Paleozoic provided a variety of niches for organisms to adapt to. While some organisms had exoskeletons, none had yet developed internal skeletons, so all the sea creatures were *invertebrates:* animals lacking an internal skeleton.

Other than the trilobites, other marine invertebrates of the early Paleozoic included reef-building creatures similar to those that build reefs in warm, shallow seas today. In fact, many of the invertebrate groups that appeared in the Cambrian have relatives alive today. Sponges, corals, sea urchins and starfish, clams, barnacles, squid, and insects are all related to the Cambrian invertebrates.

Some invertebrate fossils can be used as *guide fossils* or *index fossils:* fossils of organisms that lived in a wide variety of places but only for a short time (geologically speaking). They're useful in determining the age of rock layers they are found in.

For example, while trilobites were common throughout the 300 million years of the Paleozoic, another organism called an *archaeocyathid* existed only during the 60 million years of the Cambrian period. Archaeocyathids were reef-building organisms that lived on the sea floor and went extinct at the end of the Cambrian. Finding a trilobite fossil in a rock tells you that rock layer formed sometime during the Paleozoic. Finding an archaeocyathid fossil indicates that the rock layer containing these fossils was created during the more specific Cambrian period within the Paleozoic era.

After the archaeocyathids went extinct, other animals took over the role of reef building, including early forms of corals, sponges, and echinoderms (early ancestors to starfish and sea urchins). But reef builders were not the only creatures inhabiting the shallow seas. While it is impossible within the scope of this book to describe all the diverse Paleozoic invertebrates, I describe a few of the most fascinating and important ones next.

Swimming freely: Ammonoids and nautiloids

Ammonoids are the spiral-shelled distant relatives of modern squid that flourished during the late Paleozoic. They went extinct along with the dinosaurs at the end of the Mesozoic (see Chapter 20). Ammonoid evolution in the Paleozoic can be traced through the increasing complexity of their shells.

Ammonoid shells were spiral-shaped with individual chambers attached along sutures. The suture patterns of ammonoids became more complex through time, as illustrated in Figure 19-2. This feature allows geologists to accurately date rock layers when they find ammonoid fossils. (In Chapter 16, I explain relative dating using fossils.)

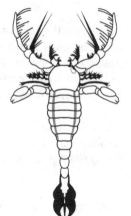

FIGURE 19-2:
Changes
in ammonoid
shell sutures
through time.

Some nautiloids appear similar to ammonoids because they have curved shells. They were both part of the larger group of mollusks. Early nautiloids had straight shells, whereas the modern living nautilus has a curved shell. None of the nautiloids developed the elaborate shell sutures seen in ammonoid fossils. A straight-shelled nautiloid is illustrated in Figure 19-3.

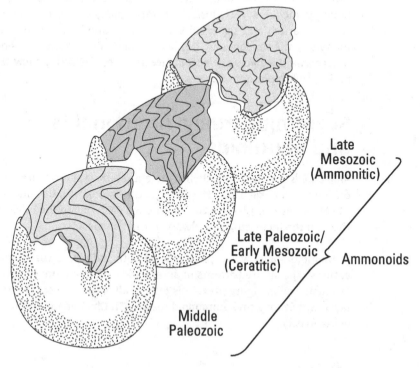

Late
Mesozoic
(Ammonitic)

Late Paleozoic/
Early Mesozoic
(Ceratitic)

Ammonoids

Middle
Paleozoic

FIGURE 19-3:
A straight-shelled
nautiloid from
the Paleozoic.

Exploring freshwater: Eurypterids

In the Silurian period, an arthropod called a *eurypterid* appeared (see Figure 19-4). Some eurypterids looked like large scorpions with frighteningly large pincers, and others were just frighteningly large. Eurypterids achieved something that most marine invertebrates of the Paleozoic did not: They successfully moved into and adapted to freshwater habitats. Fossils of eurypterids from the late Paleozoic are found in diverse habitats (freshwater and saltwater), and some have leg-like appendages that could have helped them move across short distances on land — similar to crabs today.

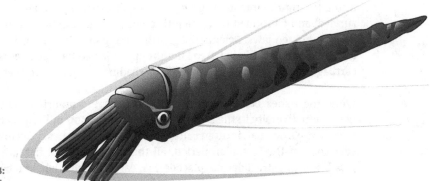

FIGURE 19-4: A eurypterid.

Spinal Tapping: Animals with Backbones

While the early Paleozoic was ruled by invertebrates, the development of skeletal features had also begun. The evolutionary story of *chordates* — animals with a nerve chord (which later includes animals with a backbone, or *vertebrates*) — is missing its earliest chapters in the fossil record because there were no hard skeletal parts to preserve.

When vertebrate fossils do show up in the fossil record, they are already full-fledged fish with backbones. And due to the presence of an internal skeleton *(endoskeleton)* and other hard parts (such as teeth and bony scales), the evolution of the fishes is a well-detailed story that I tell here.

Fish evolve body armor, teeth, and . . . legs?

The first fish were not much like the fish you see today. They had spinal cords but no jaws and were called *ostracoderms* (which means "shell skin") due to the bony plates covering them. Ostracoderms are members of the group called *agnathans* and are distant relatives of the lamprey and hagfish: two modern fish that do not have jaws or bony skin. Ostracoderms were bottom-feeders, skimming the surface of the seafloor sucking up food while keeping their eyes (located on tops of their heads) peeled for predators. They flourished through the early Paleozoic and lived alongside other fish groups evolving at the same time. Figure 19-5 is a sketch of an ostracoderm.

The earliest fish with jaws belonged to a group called *acanthodians* that no longer exist. The evolution of jaws appears to be related to the gill structure of early fish. Scientists think the frontal gill supports made of cartilage or bone may have originally become hinged to allow the gills to open wider, taking in more oxygen and also allowing the intake of more food. This feature proved advantageous to their survival and continued to develop through natural selection, resulting eventually in bony hinged jaws. Scientists are still sorting out the details but think that the acanthodians very likely led to later groups of jawed fish, such as the placoderms, cartilaginous fish, and bony fish. (Keep reading for details on these fish groups.)

While the fishes began to evolve in the early Precambrian, they reached their maximum diversity in the Devonian period (about 400 million years ago). For this reason, the Devonian is often called the "Age of the Fishes." During the 50 million year span of the Devonian period, all the major types of fish are present in the fossil record: ostracoderms, placoderms, cartilaginous fish, and bony fish.

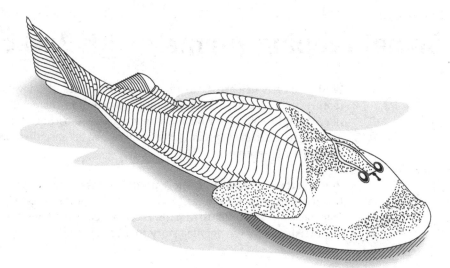

FIGURE 19-5:
An ostracoderm, the earliest fish.

Placoderms

Placoderm fish flourished in the middle to late Paleozoic but do not have any living ancestors today; by the end of the Devonian period they were extinct. The name *placoderm* means "plate-skinned" and refers to the heavily armored skin of these fish. They survived in both saltwater and freshwater habitats. Some of them, such as the *Dunkleosteus*, were frightening predators with razor-sharp teeth and body lengths of 10 to 12 meters (more than 30 feet)! The heavily armored head of a Dunkleosteus is illustrated in Figure 19-6.

Cartilaginous fish

The *cartilaginous* fish group includes modern sharks and stingrays. These creatures have jaws and teeth, but their internal skeleton is made of cartilage instead of bone. *Cartilage* is a flexible, sturdy, organic material that is found in many other animals as well; you are most familiar with it because it's what shapes your ears and nose.

Bony fish

The bony fish are by far the largest and most diverse fish group. Many of the subgroups of bony fish still exist today, and it appears that amphibians evolved from bony fish. The bony fish are separated into two groups:

>> **Ray-finned fish:** This is the type of fish you are most familiar with. These fish have fins supported by tiny bones that spread out and, when moved, propel the fish through the water. This group includes modern fish such as trout, bass, and catfish.

>> **Lobe-finned fish:** This type of fish is much rarer today — and more fascinating evolutionarily because they were the first step toward land-dwelling animals. One type of modern lobe-finned fish, the lungfish, lives in freshwater habitats such as streams or lakes and breathes through gills like other fish. However, when the water dries up, this fish can burrow into the mud and breathe through a lung-type organ until the water returns.

Among the extinct lobe-finned fish, the *crossopterygians* group is thought to have given rise to amphibians, which were the first animals to live outside water. The lobed fins of crossopterygians were muscular enough that scientists think they could have propelled these fish across short distances on land, a precursor to fully developed legs in amphibians. Other similarities between the skeletal and tooth structures of these lobe-finned fish and early amphibians are still being studied as scientists fill in the details of early land-dwelling animals.

FIGURE 19-6:
Armored head
bones of a
Dunkleosteus.

Venturing onto land: Early amphibians

In the middle Paleozoic, while the fish dominated the seas, amphibians evolved. *Amphibians* are animals that breathe air and can move comfortably outside of water but still spend the first part of their lives (as eggs and larvae) in the water. By the Devonian period, insects and plants had colonized the land, providing food resources for amphibians when they ventured out of the water.

REMEMBER

The discovery of a fossil called *Tiktaalik roseae* provided scientists with a missing link between water-dwelling and land-dwelling animals. *Tiktaalik roseae* is informally called a "fishapod" because it had characteristics of both lobe-finned fish and four-legged animals (called *tetrapods*).

Scientists think that early amphibians developed limbs to help them move around the swampy, shallow water environments of the middle Paleozoic. Amphibians became most diverse and abundant in the late Paleozoic as they spent more time outside the water. It appears that they gave rise to the next major animal group in Earth history: the reptiles.

Adapting to life on land: The reptiles

Reptiles didn't begin to dominate the earth until the Mesozoic (see Chapter 20), but they evolved and established themselves — conquering the land for vertebrates — in the late Paleozoic.

REMEMBER

In order to live on land, animals had to develop certain characteristics that allowed them to be away from water. Amphibians lived the first stage of their lives in the water and had to return to water to lay their eggs. With the appearance of an amniote egg, reptiles no longer needed water the way amphibians did. An *amniote egg* is an egg with a yolk sac inside it that provides nutrients to the developing embryo so that when it hatches from the egg, the animal is well past the larval stage and doesn't need to live in water.

Early reptiles included the *pelycosaurs*, such as the Dimetrodon fossil pictured in this book's color photo section. These animals had large fins along their backs. Paleontologists think the fins may have helped them control their body temperature. Reptiles are *cold blooded*, which means they have no internal way of warming themselves up (unlike warm-blooded animals, who regulate their own body temperature). But the pelycosaurs and their later relatives, the *therapsids*, may have begun to develop methods of body heat regulation that were precursors to the way mammals control their body temperature. I provide more information about reptiles and the evolution of early mammals in Chapter 20.

Reptiles may have been the first animals to dominate the land, but they were not the first living creatures on land. Long before amphibians ventured out of the water, plants had established themselves on land and were living alongside swarms of insects. Keep reading to find out more about what the fossil records tells scientists about life on land in the Paleozoic.

Planting Roots: Early Plant Evolution

REMEMBER

The first land plants must have started as *aquatic* (water-living) plants. In order to survive on land, plants need a supportive structure and a way to move water through their system. A material called *cellulose* in plant cell walls provides that structure, and specialized cells function to transport water throughout a plant. Plants with this special tissue are called *vascular plants.* Nonvascular plants still survive today, such as mosses and fungi, but they live in moist areas and do not grow very large because they lack vascular tissue.

During the Paleozoic, land plants experienced many important changes. By the middle Paleozoic, woody tissue (made of a material called *lignin* and much stronger than cellulose) in some plants allowed them to grow very tall. Fossils indicate that some of them were as large as modern trees, although they didn't reproduce using pollen and flowers. (These features evolved after the end of the Paleozoic, in the Mesozoic.) These early plants reproduced through *spores*, similar to modern fern species.

During the Carboniferous period (359 to 299 million years ago), ancient tree-like plants such as *Lepidodendron, Sigullaria,* and *Calamites* grew abundantly in low-lying swampy environments. These dense areas of vegetation were later pre-served and transformed into the sedimentary rocks that provide modern fossil fuel resources such as coal and oil. Figure 19-7 illustrates some of the common plants of the Carboniferous swamps.

FIGURE 19-7:
Plants common in the Carboniferous coal swamps of the Paleozoic: Lepidodendron, Sigullaria, and Calamites.

By the end of the Paleozoic, modern plant types including ferns and gymnosperms had begun to appear. Gymnosperms are plants that reproduce using pollen and seeds, but without flowers. Gymnosperms include modern plants such as *conifers*, or cone-bearing trees and cycads. Table 19-1 summarizes when these and other major steps in plant evolution occurred.

TABLE 19-1

Major Developments in Paleozoic Plant Evolution

Period	Plant Characteristic	Modern Example
Ordovician	Nonvascular	Mosses
Silurian	Vascular, land plants	Club moss
Devonian	Woody growth, leaves	Equisetum (Horsetails)
Carboniferous	Cones, seeds	Conifers

Tracking the Geologic Events of the Paleozoic

It's easy to get distracted by the abundance and diversity of life that appears and flourishes during the Paleozoic. But life and evolution are influenced by the geologic processes that are always shaping the earth's environments. The Paleozoic saw periods of intense mountain building, extensive glaciations, widespread shallow seas, and the continued buildup of material onto the continental cratons, building the continents into shapes resembling what you see today.

Constructing continents

The history of each continent is told in its rocks. Beginning with the ancient cratons that formed during the Precambrian (see Chapter 18), geologists interpret the geologic history of the continents from the rock sequences and the stories they tell.

When the Paleozoic era began, there were six major continents on Earth, none of them as large as the modern continents. These continents moved around under the influence of plate tectonics (see Part 3). Rock layers on the modern continents indicate intense periods of mountain building that occurred during the Paleozoic era as the continents crashed into each other. Each continent grew larger through the *accretion of terranes*, and mountains were built along *mobile belts*. Here's what these processes entail:

>> **Enlarging continents through accretion of terranes:** Rocky material from one continent can be added to another continent in a process called *accretion*. The foreign materials (the new rocks) have a different history than the continent they are added to and are called *terranes*. As the accretion of terranes occurs repeatedly through time, the continents grow larger and have different shapes.

>> **Building mountains along mobile belts:** When two continents collide, crustal material along the edges is forced upward (I describe the details of this process in Chapter 9). The result is elevated areas of topography — mountains — in a linear pattern parallel to the edge of the colliding plates.

Reading the rocks: Transgressions and regressions

Extensive sedimentary rocks formed in the Paleozoic era indicate times when vast, shallow seas covered the continents, depositing sandstone, shale, and limestone. These rocks are informative because sedimentary rocks, as I describe in Chapter 7, record important information about the environments in which they are deposited — particularly when the sedimentary rocks are formed by the settling of particles through water. In such cases, the sediment particles are subject to the laws of physics and gravity that still apply today. For example, a larger, heavier particle will settle out of water more quickly than a smaller, lighter particle.

When rivers or streams (described in Chapter 12) carry sediments from the continent to the sea, the sediments are deposited according to their size as the motion of the water slows down. This means that the largest, heaviest particles are deposited closest to the seashore, while the tiniest particles are carried much farther and deposited in the deep, still waters far from the shore.

The result is that the sedimentary rocks formed in the ocean have a distinct pattern to their particle sizes, with sandstone created closer to shore in shallow water and limestone created farther away in deeper water, as illustrated in Figure 19-8.

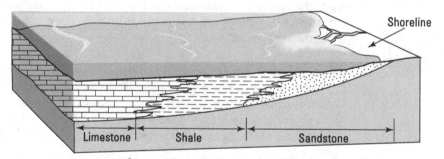

FIGURE 19-8: The pattern of sedimentary rock formation in the ocean.

Shoreline

Limestone   Shale   Sandstone

REMEMBER

By understanding how the depth of the water and distance from shore affects the type of rock formed, geologists can look at the rocks and read a story of changing sea levels.

When sea levels rise and cover more of the continent, the geologic event is called a *marine transgression.* The sedimentary rocks being deposited in one spot change from sandstone to shale to limestone as the water gets deeper at that location. This situation is illustrated in Figure 19-9.

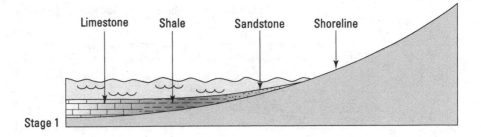

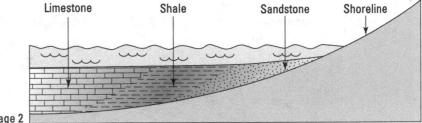

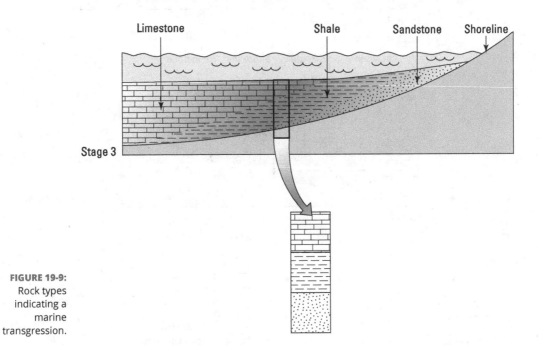

FIGURE 19-9:
Rock types
indicating a
marine
transgression.

When sea levels drop and expose more of the continent, the geologic event is called a *marine regression.* As it occurs, the type of rocks forming in one location shift from limestone to shale to sandstone as the water becomes shallower. The rock types in a marine regression are illustrated in Figure 19-10.

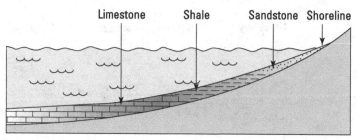

Stage 1

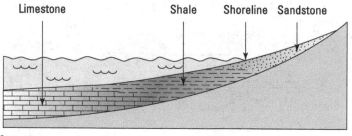

Stage 2

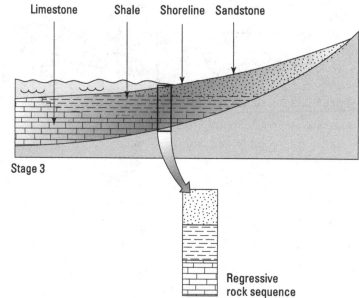

Stage 3

Regressive
rock sequence

FIGURE 19-10:
Rock types
indicating a
marine
regression.

REMEMBER

One aspect of a continent's history is found in its *cratonic sequence,* or record of marine transgressions and regressions. North America, for example, has four cratonic sequences dated to the Paleozoic. In each sequence, rocks indicate that the North American craton was covered by transgression of a shallow sea, which then regressed.

In some regions, these seas became so shallow that they dried up, leaving rocks called evaporites. *Evaporites* are minerals that form as water evaporates (see Chapters 6 and 7). In other regions, extensive reefs were built by reef-building invertebrates that later became geologic deposits of limestone.

Fossilizing carbon fuels

Concern about modern climate change has people talking about reducing the use of *fossil* or *carbon* fuels. Both terms refer to coal, oil, and natural gas resources, many of which are found in Carboniferous rock layers created during the late Paleozoic. The abundant plant life in the Carboniferous period left its carbon remains to form geologic deposits in rock layers of coal. In fact, geologists describe these mid- to late-Paleozoic environments as *coal swamps*. (Some coal swamp deposits were formed during the following period, the Mesozoic, which I describe in Chapter 20.)

REMEMBER

A rock sequence common in the Carboniferous period, especially the Pennsylvanian period, is called a *cyclothem*. Cyclothem rock layers indicate a transition between marine and nonmarine environments, similar to what is observed today at a low-lying river delta such as the Mississippi River. These regions are now (and were in the Carboniferous period) full of thick, swampy vegetation. As these plant materials die, become buried by sediments, and accumulate over time, they turn into coal beds.

The cyclothems of the Carboniferous are so widespread that geologists are still seeking answers to how they were formed because it seems unreasonable that so much of the earth was covered in swampy, transitional marine environments.

Pangaea, the most super of supercontinents

By the end of the Paleozoic, all the major continents on Earth came together, forming a single supercontinent. Pangaea is the most super of supercontinents because there is no evidence that ever before — or ever since — all the major landmasses formed a single continent. In Chapter 8, I explain that the remains of Gondwana, the southern landmass of Pangaea, led early geologists to ask the questions that eventually led to the theory of plate tectonics.

On the supercontinent of Pangaea in the early Mesozoic is when dinosaurs began to evolve. To find out what happens when the supercontinent Pangaea breaks apart in the Mesozoic era, be sure to read Chapter 20.

Chapter **20**

Mesozoic World: When Dinosaurs Dominated

A pproximately 250 million years ago (251 million to be exact), reptiles took over from the amphibians to rule the continents. For a span of almost 200 million years — the Mesozoic era — reptiles evolved to fill every habitat on the planet. From the equator to the poles, on land and in the sea, reptiles ruled.

The Mesozoic era extends from 251 to 65.5 million years ago and includes the Triassic (251–201 million years ago), Jurassic (200–145 million years ago), and Cretaceous (146–65.5 million years ago) periods. It is also called the *Age of Reptiles* or the *Age of Dinosaurs.* By the beginning of the Mesozoic, reptiles were diverse and numerous enough to take the place of many amphibian and invertebrate groups that went extinct at the end of the Paleozoic era, the topic of Chapter 19. (In Chapter 22, I describe the multiple extinction events in Earth's history.)

In this chapter, I explain what happened to the earth's continents between 251 and 65.5 million years ago. I also explain the evolution of the reptiles and provide some details on dinosaurs and other Mesozoic reptiles that took to the skies or seas. Finally, I describe how during this time mammals appeared and began their evolutionary journey to eventual domination following the extinction of the dinosaurs at the end of the Mesozoic.

Driving Pangaea Apart at the Seams

Geologists have a good record of rock-forming events and processes during the Mesozoic relative to earlier eras. The details of Earth's geologic history become more detailed in more recent rock layers because the rocks have not been as deformed as older rocks (like those of the Paleozoic and Precambrian eras). The Mesozoic rock layers provide scientists with a huge amount of information about the development of the modern continents.

One continent becomes many

When the Paleozoic era ended around 250 million years ago, the earth's land-masses were connected, forming a single supercontinent called Pangaea (which I introduce in Chapter 19). Pangaea was surrounded by a single large mass of water called the *Panthalassa Ocean*. The modern continents were arranged to form Pangaea as illustrated in Figure 20-1, stretching from pole to pole and centered on the equator.

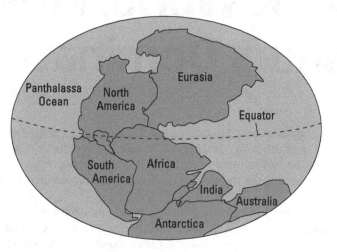

FIGURE 20-1:
The arrangement of modern continents when they formed Pangaea.

As the Mesozoic era began, large portions of the southern part of Pangaea (at the South Pole) were still covered in the glaciers that developed during the cold climate of the previous era.

Fifty million years later, two sections of Pangaea — *Laurasia* and *Gondwana* — began to separate. These two sections are illustrated in Chapter 8 (refer to Figure 8-6). In that chapter, I explain how the geological evidence of Gondwana is what inspired the geologists who developed early ideas about plate tectonics theory.

Another 50 million years later, South America and Africa begin to spread apart, creating what is now the Atlantic Ocean. By the end of the Mesozoic era, the continents looked very much like they do today, except that India had not yet connected to Asia and Greenland was still attached to Europe.

Influencing global climate

For the span of human history, the continents have always been in the same place, so it may never have crossed your mind to wonder what the weather in your region would be like if the Himalayan Mountains were near Australia instead of between India and China. But mountains and other features of the earth's surface shape the patterns of wind and water that circulate weather around the planet. The size of a continent, size of an ocean, distance from the equator, and elevation of the continent all play a role in global weather patterns today, just as they have in the past.

REMEMBER

The breakup of Pangaea dramatically affected the global climate and weather patterns, creating new environments across the land and beneath the seas. The changes in climate caused by the breakup of Pangaea are recorded in the Mesozoic rocks. Patterns of two rock types in particular indicate changes in the global climate patterns during the Mesozoic era.

>> **Evaporites:** Rocks called *evaporites* are created when an area covered by water dries out. As the water evaporates, minerals form from the elements that were dissolved in the water. Mesozoic-aged evaporites are found in regions that were near the equator between 250 and 65 million years ago, suggesting that the climate near the equator was especially warm and dry. (This is different from conditions near the equator today, which are warm and wet.)

>> **Coal:** I explain in Chapter 19 that coal is formed by the accumulation of organic matter, such as plants, in a warm, tropical region. Prior to the Mesozoic (during the Carboniferous period, which I describe in Chapter 19), coal deposits were formed in regions near the equator. But during the Mesozoic, coal layers began to form much closer to the poles, suggesting that after the Paleozoic glaciers melted, temperatures at the North and South Pole were fairly warm.

Scientists generally accept that the Mesozoic climate was moderate and warm. Some scientists think that the movement of landmasses away from the equator toward the poles began to create a temperature gradient similar to, but not as extreme as, the one that exists today. (A *temperature gradient* describes how the temperatures get cooler as you move from the equator toward the poles.) However, other scientists think that during the Jurassic period (the warmest part of the Mesozoic), average temperatures were almost the same all over the globe.

Creating the mountains of North America

All over the world, continental landmasses were changing in size and shape as the continents of Pangaea moved apart. I focus here on just the changes in North America.

The eastern portions of North America, specifically the Appalachian Mountains, had already formed by the start of the Mesozoic era. During the Mesozoic, North America experienced a series of mountain-building episodes, or *orogenies*, along its western coast. The first of these is called the *Cordilleran orogeny* and created the Sierra Nevada and Rocky Mountains. As the North American continental plate and the ancient Farallon Plate moved toward each other, the Farallon Plate subducted (a process I explain in Chapter 9), and volcanoes were formed along the plate boundary. (Only a small piece of the Farallon Plate now remains and is called the *Juan de Fuca Plate*. It's located off the northwest coast of the United States.) These volcanoes formed an island arc that was eventually *accreted* (added) to the western coast of the North American continent.

Meanwhile, the previously formed Appalachian Mountains were eroding, moving sediments westward to settle across the continent at the bottom of a shallow sea called the *Sundance Sea* that covered what is now the central part of North America.

By the end of the Mesozoic, the Sundance Sea was a waterway crossing the continent from north to south called the *intercontinental seaway* (see Figure 20-2). The sediments deposited in the intercontinental seaway were washed down, forming what is now called the Gulf of Mexico.

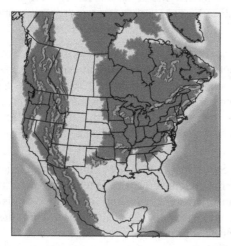

FIGURE 20-2:
North America during the Mesozoic era.

Repopulating the Seas after Extinction

The breakup of Pangaea and resulting changes in global environments were driving forces in the evolutionary story of the Mesozoic. The previous era ended with a mass extinction that wiped out most of the marine invertebrate populations. A *mass extinction* is when numerous species and even groups of species are killed off. The end-Permian extinction that preceded the Mesozoic is the largest mass extinction event but only one of many mass extinction events in Earth's history; I describe all of them in Chapter 22.

At the beginning of the Mesozoic, many of the species that had been filling the oceans were no longer around. Separate landmasses and new oceans provided new habitats to be filled by creatures with the most advantageous characteristics.

REMEMBER

When organisms move into new habitats, they may evolve different characteristics to best fit the new habitat. Scientists call this process *adaptive radiation.* The result is that a group of animals that may have previously been one species restricted to a specific environment or region now has multiple species filling all different environments and regions, each one with characteristics uniquely suited to its habitat. Adaptive radiation leads to *diversification* as the species develop new characteristics.

In the early Mesozoic, life in the oceans rebounded from the end-Permian extinction in some interesting ways. For example, the number of different types of burrowing animals increased. Whereas previously only soft-bodied animals burrowed into the sediments of the sea floor, fossils indicate that in the Mesozoic, animals with shells also burrowed into the sediments. Scientists think it is likely that these animals evolved this new way of life to hide from predators that even their shells couldn't protect them from.

Single-celled organisms called *planktonic foraminifera* first appeared — and then evolved increasing diversity — during the Mesozoic. Foraminifera are tiny animals with only one cell that build mineral shells. *Benthic* foraminifera live at the bottom of the ocean and have been present since the Cambrian period at the beginning of the Paleozoic (approximately 540 million years ago). Planktonic foraminifera float near the surface of the ocean and first appeared in the Jurassic. They were so diverse and abundant during the Cretaceous period of the late Mesozoic that they are used as guide fossils. (I define guide fossils in Chapter 19.) The shapes of planktonic foraminifera are illustrated in Figure 20-3.

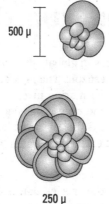

FIGURE 20-3:
Planktonic
foraminifera.

500 μ

250 μ

The fishes continued to diversify and evolve new species throughout the Mesozoic. Fish evolution is so detailed and specific that it's easier to categorize the fish species as *primitive, intermediate,* and *advanced.* The most obvious difference is that the advanced group had a completely bony skeleton, unlike the earlier groups whose skeletons were largely made of softer cartilage material (see Chapter 19). By the end of the Mesozoic, the most complex fish group was the *teleosts.* Teleosts are ray-finned fish with bony skeletons that inhabit fresh and salt water. They are still the most diverse and abundant vertebrate animal group on the planet!

The Symbiosis of Flowers

The changes seen in the fossil record for plants between 251 and 65.5 million years ago lay the foundation for the majority of plant species on Earth today. As the continents of Pangaea broke apart and moved around, new habitats were created on land as well as in the sea. In response to these new habitats, it appears that plants evolved two specific characteristics that proved very advantageous:

» **Enclosed seeds:** Seeds that are *enclosed* or protected by a shell can travel longer distances and survive through poor environmental conditions longer than unenclosed (and therefore unprotected) seeds. This means that they can wait until conditions are just right to begin growing a new plant.

» **Flowers:** Flowers provide food to insects. In exchange, the insects carry pollen from one plant to the next, acting as the fertilizing agents of plant reproduction. This gives flowering plants a huge advantage, spreading their genetic information and reproductive success far and wide across the landscape.

EVOLVING TOGETHER: FLOWERS AND INSECTS

The evolution of flowering plants and the evolution of insects are closely intertwined, so much so that scientists call it *co-evolution*. This symbiotic relationship benefits both of them: Flowers provide food for the insect, and when the insect collects that food, it unknowingly carries the plant pollen from one flower to the next. In this way the insect is an important part of the plant's reproductive system. This relationship began in the Mesozoic. Once they were entwined, the fates (and evolutionary characteristics) of flowers and insects continued to shape each other for millions of years.

I'm sure you've noticed the wide range of flower diversity: They are different shapes, sizes, colors, and scents, and they even bloom at different times of the year. Each of these characteristics evolved to entice a particular insect into a reproductive dance with a particular flower. For example, bees cannot see the color red, so bees are usually attracted to blue or yellow flowers. When a bee lands on one of these flowers, the pollen, which must be carried to another flower in order to produce a seed, sticks to the legs of the bee and flies with it to the next flower. A close look at the shape, size, and features of some pollen grains makes it clear that they have evolved to hitch a ride with visiting insects and travel to the next flower.

A more spectacular example is the *stinking corpse lily*. Yes, *corpse*, which means dead creature. The stinking corpse lily is a large, odorless flower most of the time. But when it is ready to reproduce it blooms, it sends out a strong scent — the scent of something dead and rotting. This scent attracts flies, who love a dead, rotting corpse for dinner. When the flies swarm onto the flower, they pick up its pollen and carry it with them to the next stinking corpse lily they find.

More recently, this type of co-evolution has also been seen in birds and flowers — for example, in hummingbirds and the large, red flowers they drink nectar from. Unlike bees, they can see the color red, and the beak of a hummingbird is long and thin, perfect for dipping into the bright red bell-shaped hibiscus flower.

The development of flowers, in particular, is a turning point in not only plant evolution but also insect evolution. In fact, the evolution of flowers and insects is so closely related that it is considered *co-evolution*. Co-evolution is a type of *symbiosis* or relationship between two organisms. In this case, both the plant and the insect benefit from their relationship. Today some plants and insects have such closely co-evolved characteristics that only a certain species of insect can pollinate a certain species of plant.

Recognizing All the Mesozoic Reptiles

The front-page story of Mesozoic life on Earth is, quite obviously, the reptiles — specifically the dinosaurs. But contrary to popular belief, not all ancient reptiles were dinosaurs. Dinosaurs are only one branch on a family tree that includes modern reptiles such as turtles, snakes, and lizards, along with surviving ancient reptiles such as crocodiles (which evolved 200 million years ago and are still around) and others.

At the end of the Paleozoic, animals called *protorothyrids*, or *stem reptiles*, were present. (In Chapter 19, I briefly explain the relationship of the Paleozoic reptiles.) Scientists call them stem reptiles because they form the base or stem of the reptile family tree, from which all the other reptiles (including dinosaurs) evolved. By the beginning of the Mesozoic, the ancestors of all ancient and modern reptiles had evolved from the protorothyrids. This includes the *archosaur* (ancestor to the dinosaurs), crocodiles, and — much later — birds. The relationships between ancient and modern reptiles are illustrated in Figure 20-4.

As you can see from the simplified diagram in Figure 20-4, modern lizards, snakes, tuataras, and turtles are only distantly related to the swimming reptiles, flying reptiles, and dinosaurs. In this section, I describe the characteristics of the ancient and extinct reptiles.

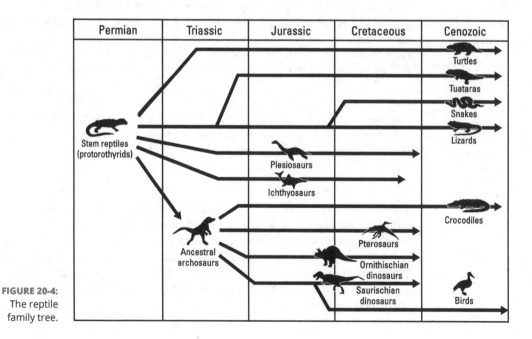

FIGURE 20-4: The reptile family tree.

Swimming in ancient seas

For millions of years in the Paleozoic, early reptiles populated the land. By 251 million years ago, a few of them had returned to the seas. A few groups of reptiles evolved to survive in the ancient seas of the Mesozoic earth:

>> **Ichthyosaurs:** *Ichthyosaurs* were large, fish-like reptiles that looked somewhat like modern dolphins. They gave birth to live young instead of laying eggs like most reptiles. They were predators, at the top of the marine food chain in the Mesozoic, eating fish and other marine creatures, including cephalopods (see Chapter 19).

>> **Plesiosaurs:** The *plesiosaurs* were fish-eating reptiles. Some of them had short necks, and others had very long necks, a tail, and four paddle-like feet. They ranged in size from 4 meters to up to almost 15 meters (13 to almost 50 feet).

>> **Mosasaurs:** *Mosasaurs* evolved toward the end of the Mesozoic, during the Cretaceous period. They resembled large lizards with paddle-shaped limbs. Scientists now think that mosasaurs are closely related to modern snakes, which also evolved from early lizards.

Taking to the skies: Pterosaurs

REMEMBER

The first vertebrate animals to fly were Mesozoic reptiles called *pterosaurs*. Most of the pterosaurs were small, about the size of the birds you might see in your backyard. (They were not birds though!) Some scientists have suggested that the largest pterosaurs, like the *pteranodon* illustrated in Figure 20-5, were probably too large to fly using their relatively small and weak wing structure.

FIGURE 20-5:
A flying reptile of the Mesozoic, the pteranodon.

Pterosaurs had hollow bones like modern birds, and some may have had hairy or feathery coverings. They ate fish and, unlike most reptiles, were warm-blooded.

Flocking together

Fossils of reptiles and early birdlike animals from the Mesozoic era have led scientists to understand the relationship between reptiles and birds. Modern reptiles and birds share common features such as egg laying and certain skeletal characteristics that indicate they probably have a common ancestor. With fossil finds, such as *Archaeopteryx*, links between modern birds and ancient reptiles, in particular dinosaurs, have become clearer. Keep reading to find out the details of the evolutionary path from reptiles, to dinosaurs, to modern birds.

Climbing the Dinosaur Family Tree

Clearly the most dominant life forms on the Mesozoic earth were the large (and small) land-dwelling reptiles: the dinosaurs. Movies often portray dinosaurs as giant, scary monsters. In fact, dinosaurs were widely variable in their size, eating habits, and living patterns. Consider mammals today; while a tiger may be a large, ferocious predator, the squirrel in your backyard isn't so frightening. During the Mesozoic, dinosaurs filled both these roles — and everything in between.

REMEMBER

The dinosaur family tree has two main branches: *Ornithischia* and *Saurischia*. The division in the branches was originally based on the skeletal structure of the pelvic bones (the ilium, ischium, and pubis) but further study of fossils has expanded the defining traits of these two groups to include other features of the skeleton. In this section, I describe the identifying characteristics of these two main groups, as well as particular dinosaurs and where they fit into the dinosaur family tree.

Branching out: Ornithischia and Saurischia

Fossilized dinosaur bones, specifically their hip and pelvis bones, gave scientists a simple way to sort out who was related to whom. With more modern studies, the use of cladistics (I explain cladistics in Chapter 17) has enhanced our understanding of evolutionary relationships, but the differences observed by early paleontologists still hold true.

WARMING THEMSELVES FROM WITHIN: WERE DINOSAURS WARM-BLOODED?

For a long time, scientists accepted that dinosaurs were reptiles and assumed that all dinosaurs were exothermic (cold-blooded) the way modern reptiles are. But as scientists have continued to examine the fossils of ancient reptiles and dinosaurs, they are no longer certain that the answer to this question is so simple. Some of the arguments scientists make for dinosaur endothermy include these:

- Birds are direct descendants of dinosaurs, and birds are endothermic.

- The larger dinosaurs (like a Brachiosaurus) would have needed an endothermic circulatory system to pump blood up their long necks.

- Fossils of dinosaurs are found all over the globe, including the Arctic and Antarctic regions, suggesting they had some way to keep warm in these cooler regions.

- Evidence indicates that some dinosaurs had feathers or fur, features that are associated with modern endothermic animals.

These arguments do not provide irrefutable proof for dinosaur endothermy. Opposing scientists point out that a large dinosaur such as the Brachiosaurus would not need an advanced circulatory system if it carried its neck straight (horizontal to the ground) rather than up (like a giraffe). Others claim that the climate during the Mesozoic was warm enough, even at the North and South Pole, that endothermism may not have been required for dinosaurs to live there. With only bones and partial skeletons to work with, scientists continue to search for evidence to support their multiple hypotheses and determine if some dinosaurs were indeed warm-blooded.

Horned faces and armor: Ornithischian dinosaurs

Dinosaurs on the *Ornithischia* branch are described as having birdlike pelvic bone structure, which is illustrated on the left in Figure 20-6.

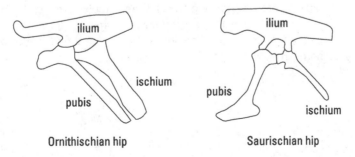

FIGURE 20-6: Bird- and lizard-like hip structure of dinosaurs.

Ornithischian hip

Saurischian hip

REMEMBER

The ornithischians include dinosaurs that were *bipedal*, or walking on two legs, as well as *quadrupedal*, walking on four legs. As far as scientists can determine, all the ornithischian dinosaurs were vegetarians, or *herbivores*, eating only plants (this conclusion is based on the teeth and jaw structure). Among ornithischians you find some of the more dramatic expressions of body adornment. For example, Thyreophora includes the *Stegosaurus* and *Ankylosaurus* as well as related dinosaurs with bony plates or *osteoderms* along their backs. And Marginocephalia includes *Ceratopsia* and *Pachycephalosaurus*, both dinosaur groups with large, thick skull ornaments, as well as Ornithopoda or duck-billed dinosaurs. The Ornithischia group includes the following dinosaurs, which are illustrated in Figure 20-7:

>> **Ankylosaurs:** The most heavily armored dinosaurs were the *ankylosaurs.* Its entire back, parts of its legs, and top of its head were covered in a thick, protective, bony armor. It also had a large club-tail that could be used as a defensive weapon. This armoring meant that the ankylosaurs could weigh up to 5 tons! Because they carried this much weight, scientists estimate that they could travel only about 5 to 8 miles an hour. Closely related to the ankylosaur is the *nodosaur*, which lacked the clubbed tail but instead had large, hornlike spines growing out from its shoulder.

>> **Stegosaurs:** *Stegosaurs* were quadrupedal and included the Stegosaurus from the Jurassic period, approximately 150 to 200 million years ago. Stegosaurs had large plate-like spines along their backs and spiked tails. Although the spiked tails were probably for defense, scientists are still figuring out the purpose the plates along their back served. Current hypotheses include social display to others within and without their species, protection from predators and body temperature regulation.

>> **Ceratopsians:** The *ceratopsian* dinosaurs had sharp teeth for cutting down vegetation and large horns, presumably for defense, on their heads. The most well-known ceratopsian dinosaur is the *Triceratops.* Triceratops fossils are most common at the end of the Mesozoic, around 70 million years ago. However, ceratopsians are a very diverse group and scientists believe the different species displayed different types of head ornaments (or *frills*). They were also quadrupedal and traveled in large herds.

>> **Pachycephalosaurs:** The *pachycephalosaurs*, like the ceratopsians, have a large, bony head, but unlike the ceratopsian frill, the pachycephalosaurs grew a thick skull cap right over the top of their heads. Studies of the stress fractures in fossil pachycephalosaur skulls suggest that they smashed their heads together — perhaps like modern rams who fight by running at one another and tangling horns. They were bipedal and herbivores.

>> **Ornithopods:** The *ornithopod* dinosaurs were bipedal, with front legs that were developed enough to occasionally support them as quadrupeds. This group includes the duck-billed dinosaurs, or *hadrosaurs.* The hadrosaurs (such as the *Hypocrosaurus*) had a distinctive skull that in some species extended into a bony crest. Scientists have yet to figure out what purpose (if any) these elaborate bony crests served, but current hypotheses include social signaling or even making sounds through the hollow crest to communicate.

a) Ankylosaurus

b) Stegosaurus

c) Triceratops

d) Hadrosaurus

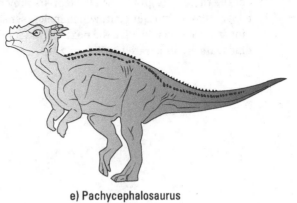

e) Pachycephalosaurus

FIGURE 20-7:
Ornithischian
dinosaurs.

Long necks and meat eaters: Saurischian dinosaurs

The *saurischian* dinosaurs had lizard-like hip structure, illustrated in Figure 20-6 on the right. This group had two distinct branches of dinosaurs, or sub-orders: the *sauropods* and *theropods.*

The sauropods were most common in the earlier part of the Mesozoic. They were massive creatures, possibly the largest land animals ever to exist, and included the *Brachiosaurus, Diplodocus,* and *Apatosaurus.* These creatures walked on four legs, ate plants, and moved in herds. The largest of the sauropod fossils are found in South America where evidence for extensive nesting by these animals has also been uncovered. Some evidence even suggests that sauropod species laid eggs near hydrothermal features so that the heat and moisture benefitted the hatching of the eggs.

REMEMBER

The theropod group includes that most famous dinosaur: the *Tyrannosaurus rex,* or T-rex. Another theropod made famous by the movie *Jurassic Park* is the *velociraptor.* Most theropods were bipedal and carnivores. The smaller theropods likely hunted in packs. However, within the group there is quite a lot of diversity! One group of theropods appears to have lived both in and out of the water. The *Spinosaurus* had a long, narrow snout; fossil evidence shows it ate fish and had tall spines along its back. Scientists are still trying to determine whether the spines supported a thin, sail-like structure (like the dimetrodon illustrated in this book's color photo section) or maybe a large back hump.

Recent discoveries of many different theropod fossils have confirmed that feathers and *proto-feathers* (less developed than modern feathers) are present throughout the theropods group. What scientists previously considered a bird trait is now apparent as a dinosaur trait, which makes sense when you consider how birds and dinosaurs are related. (See the next section for details.) Some common saurischian dinosaurs are illustrated in Figure 20-8.

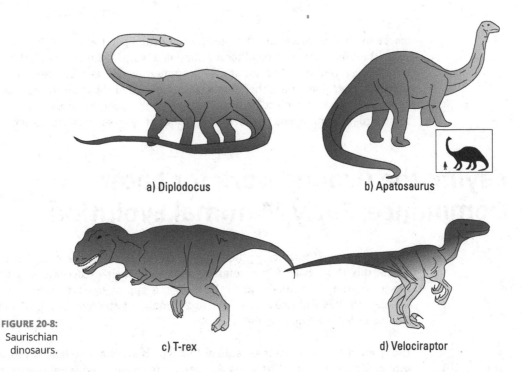

FIGURE 20-8: Saurischian dinosaurs.

a) Diplodocus

b) Apatosaurus

c) T-rex

d) Velociraptor

Flocking Together: The Evolutionary Road to Birds

One group of theropods in particular has made news for not actually being extinct! Although scientists long considered all dinosaurs extinct, fossil finds in the last few decades have rewritten that story. Starting with *archaeopteryx*, scientists began to recognize that early bird fossils shared many skeletal features with the dinosaurs, such as the size and position of ankle bones. It is now accepted that one line of theropod dinosaurs in the late Mesozoic evolved into birds — and that some early bird species coexisted with dinosaurs, and of course, survived them after the large dinosaur destroying mass extinction 65 million years ago.

One of the major questions in studying the dinosaur-to-bird evolutionary path is which came first: the feathers or the flying? Although scientists long accepted feathers as being made for flight, with the presence of feathers in obviously non-flying dinosaurs, and even proto-feathers in groups like the ceratopsians, scientist have had to accept that feathers came long before flight. So if feathers aren't for flight, what are they for? Maybe insulation, maybe as sensory tools, maybe to display coloration, the questions are still open for debate. But in recent years the understanding of dinosaurs as giant scaly-skinned lizards has been transformed, and now it's not at all uncommon to see reconstructed dinosaurs sporting colorful and patterned feathers. And not all of these reconstructions are imaginary! Modern molecular paleontology leads scientists to understand that some bird-line therapods had stripes, splashes of red, and even iridescence.

Laying the Groundwork for Later Dominance: Early Mammal Evolution

Throughout the Mesozoic, reptiles dominated the land and the seas. But it was during this time that early mammals, or at least their predecessors, began to appear. A group of animals called the *therapsids* appear to have bridged the gap between reptiles and mammals prior to 251 million years ago. One group of the therapsids led the way to mammals.

REMEMBER

The *cynodonts* were a therapsid animal that had many characteristics in common with modern mammals. These mammalian traits include some very specific skeletal features of the head, jaw, and ear. Other traits exhibited by cynodonts that are linked to mammals are an outer covering of skin instead of scales and indications that they were able to regulate their body temperature (they were *endothermic*). Many of these seemingly subtle and detailed changes are documented with fossil evidence illustrating multiple stages between the reptilian therapsid traits and the mammalian cynodont traits.

The early mammals evolved and diversified during the late Mesozoic, but I cover their evolutionary history in more detail in Chapter 21, where I discuss the Cenozoic period, also known as the Age of Mammals.

Chapter **21**

The Cenozoic Era: Mammals Take Over

The Cenozoic era of Earth's geologic history hasn't ended yet. It began 65.5 million years ago and continues today. Compared to previous geologic eras, the Cenozoic (so far) is relatively short. But because it's the most recent era, scientists have access to a huge amount of geologic and fossil evidence documenting its events.

Some geologists separate the Cenozoic era into two periods: the Tertiary (from 65.5 to 2.6 million years ago) and the Quaternary (from 2.6 million years ago to the present). However, the Cenozoic is more commonly separated into three periods: the Paleogene (65.5 to 23 million years ago), Neogene (23 to 2.6 million years ago), and Quaternary (from 2.6 million years ago to the present).

In this chapter, I explain a few of the major geologic events of the Cenozoic in North America and describe the fossil history of the rise of mammals. Entering the Cenozoic era, you enter the Age of Mammals and (eventually) humans.

Putting Continents in Their Proper (Okay, Current) Places

At the beginning of the Cenozoic era, the major continents were pretty much positioned the way they are today, with the exception of parts of Europe, Asia, and the Indian subcontinent. The major mountain ranges of the modern world were formed during the last 65 million years and are still being built today by the movement of crustal plates on Earth's surface. I discuss the origin of some of these mountain ranges next.

Creating modern geography

So much evidence for Cenozoic geologic processes is available to geologists on the surface of the earth that it could fill multiple books. In this section, I describe only two of the major mountain-building regions: the Alpine-Himalayan belt and the Circum-Pacific belt. I also explain how some of the major features of the modern North American continent have evolved over the last 60 million years.

The Alpine-Himalayan belt

The *Alpine-Himalayan orogenic belt* extends from the Strait of Gibraltar in the western Mediterranean, across the Middle East, through Turkey, and into Asia, where India and China meet at the Himalayan mountain range. This belt is illustrated in Figure 21-1.

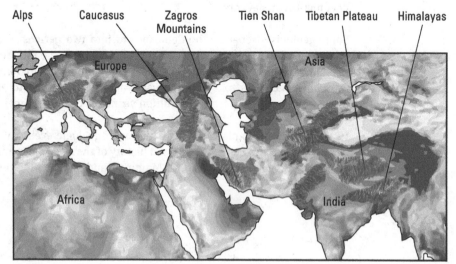

FIGURE 21-1: The Alpine-Himalayan orogenic belt.

The European Alps began forming during the middle of the Mesozoic era (see Chapter 20), but things really picked up during the Cenozoic. As the African continental plate moved northward, it crushed smaller plates into the European continent, forming the Alps. This northward motion of the African Plate continues today, compressing and deforming the rocks along the southern portion of Europe and slowly closing off the Mediterranean Sea, which used to be a much larger ocean between Africa and Europe.

Farther east, the Arabian Plate moved northward into Turkey, forming the Taurus Mountains. Today the Arabian Plate continues to move toward Asia, causing earthquakes in the region. (Check out Chapter 10 for details on plate movements and earthquakes.)

What is now the Indian subcontinent was a separate small continent at the beginning of the Cenozoic. Over a period of 55 million years, it has moved northward from the equator up into the Asian continent. Where the Indian Plate and the Asian Plate collide is the Himalayan Mountains, the most elevated portion of Earth's crust, reaching 8,848 meters (29,000 feet and still growing!) above sea level at the top of Mt. Everest, the highest mountain in the world.

The Circum-Pacific belt

The *Circum-Pacific orogenic belt* runs from the southern tip of South America, up its western coast, along western North America, across the top of the Pacific Ocean, and then down through Japan and the islands of Southeast Asia. The Circum-Pacific orogenic belt is illustrated in Figure 21-2.

This region of the earth is also called the *Ring of Fire* because it is dominated by volcanoes produced as the Pacific Plate subducts under other plates on nearly every edge. The Circum-Pacific orogenic belt is extremely active today, creating volcanic island arcs and continental arcs through subduction processes. (I describe these features and processes in detail in Chapter 10.)

Consuming the Farallon Plate

In Chapter 20, I explain that the North American Plate and Farallon Plate moved toward each other during the Mesozoic era, building the Rocky Mountains through compression, deformation, and uplift of the earth's crust. At the beginning of the Cenozoic, the Farallon Plate continued to subduct beneath the North American Plate. However, instead of diving steeply downward, it moved at less of an angle, sliding under the North American plate. The result is evident in the changing patterns of mountain building across North America. Whereas the Mesozoic episodes of mountain building activity produced volcanic mountains along the western coast of North America, the early Cenozoic mountains were formed farther inland by the lifting and deforming of the crust into the central Rocky Mountains, stretching from Colorado north into Canada. These and other features of North America are illustrated in Figure 21-3.

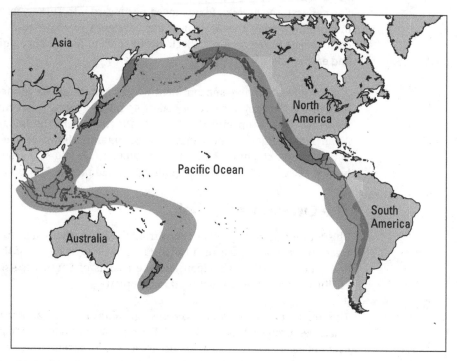

FIGURE 21-2:
The Circum-
Pacific belt, called
the Ring of Fire.

By a little more than 40 million years ago, the mountain building pattern in western North America shifted back to volcanic activity along the coast. The Cascade Mountains, illustrated in Figure 21-3, stretch from northern California through Oregon and Washington and into southern British Columbia, Canada. The Cascades are continental volcanic arcs produced by the subduction and melting of what little remains of the Farallon Plate. Only a small portion is left, and it is now called the *Juan de Fuca Plate*. It continues to move toward and under the North American plate along the northwestern United States, causing volcanic activity and earthquakes along the west coast.

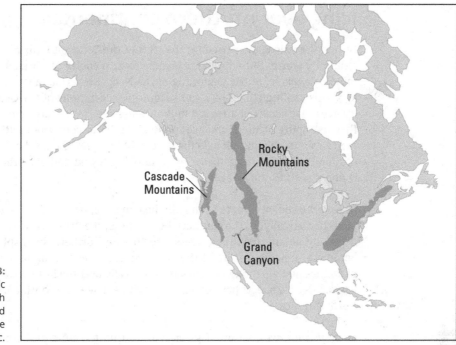

FIGURE 21-3:
Geographic
features of North
America formed
during the
Cenozoic.

Carving the Grand Canyon with uplift

Possibly the most famous geologic feature of modern North America is the Grand Canyon. You can find a photo of the Grand Canyon in this book's color photo section, and its location is indicated in Figure 21-3.

The sedimentary rock layers of the Grand Canyon were formed over a period of at least 550 million years (though many layers that should date to the last 250 million years are missing, probably due to erosion). The canyon itself could only have been carved long after the sediments that form the rocks were laid down, compressed, and *lithified* (turned to stone).

Exactly how the deep canyon has formed is still being explored by many geologists. Most accept that the removal of sediments by the Colorado River plays a large role. However, the rate of river flow and sediment removal doesn't add up to the enormous amount of eroded material from the canyon. Therefore, some scientists wonder if another process is at play. What they have found is that this region of southwestern North America is also experiencing a slow and gentle uplift (relative to other regions of the American west). The uplift began during the Neogene period of the Cenozoic, lifting the Colorado Plateau region more than 1,500 meters (more than 4,900 feet) above sea level. The combination of slow uplift and stream erosion together could explain how such deep canyons have formed in such a relatively short time (geologically speaking, of course).

Icing over northern continents

During the Paleogene period (at the start of the Cenozoic), the global climate was still fairly warm. But a dramatic change toward cooler conditions occurred in the middle Cenozoic, at the beginning of the Neogene period, about 23 million years ago. This cooling climate led to the multiple glaciations of the Quaternary period ice ages that occurred after 2.8 million years ago. Some scientists have suggested that the uplift of the Himalayan Plateau as India connected with Asia played a major role in shifting global climate conditions (see the sidebar "Causing global cooling"). Multiple hypotheses are still being tested and explored to explain the changes seen as the Neogene began.

During the cooler climate of the middle Cenozoic, layers of ice began to cover the poles periodically. In the southern hemisphere, the Antarctic ice sheet was established by about 10 million years ago. In the northern hemisphere, ice didn't become a geologic feature of the Cenozoic until the beginning of the Quaternary period, around 2.8 million years ago. Periodically over the last two and a half million years, large ice sheets have extended southward over the continents of Europe and North America.

Each *glacial stage* (period of glaciation) during the Quaternary period lasted for thousands of years. In between the glaciations were periods called *interglacials* when the ice cover shrank. The alternation of glacial stages and interglacials over hundreds of thousands of years has been linked by scientists to patterns of the earth's orbit, rotation, and distance from the sun. (I describe these cycles and other features of glaciers and glacial geology in Chapter 13.)

The modern landscapes of North America and Europe were shaped by the repeated growing and shrinking of ice sheets. As the mass of ice grew, it moved farther south, eroded rocks and sediments from the surface of the earth, pushed these earth materials southward, and deposited them as moraines and other glacial features I describe in Chapter 13.

The cycle of Neogene and Quaternary glacials and interglacials shaped the recent evolutionary story of mammals, including man.

Entering the Age of Mammals

Mammals did not suddenly appear in the Cenozoic era. By 65 million years ago, mammals had been living side by side with the reptiles and dinosaurs for almost 150 million years. At the beginning of the Cenozoic, however, conditions were right for mammals to take over as the dominant animals.

Mammals, like reptiles, are vertebrate animals, with an internal skeletal structure. But mammals have certain characteristics that separate them from reptiles. I list a few of the most easily observed characteristics here:

>> **Mammals have differentiated teeth.** This means that within the jawbone of a mammal, the teeth are shaped differently to perform different tasks. For example, in humans the front teeth (the *incisors*) are for cutting, whereas the back teeth (the *molars*) are for grinding.

>> **Mammals are endothermic.** Being *endothermic* means an animal can regulate its own body temperature, to warm up or cool down as needed. Reptiles do not have this ability (they are *exothermic*), which is why you often see snakes or lizards sunning themselves on rocks to warm up.

>> **Most mammals give birth to live offspring.** Unlike the reptiles, who lay eggs, mammals birth live young that must be taken care of for at least a little while before they can fend for themselves. (I explain the exception to this characteristic — the monotreme mammals — in a moment.)

>> **Mammals produce milk to feed their young.** Mammalian animals have *mammary glands* that produce milk intended to feed their offspring.

>> **Mammals have fur or hair covering their bodies.** Modern mammals live in such a wide variety of environments that you may not realize they all have some hair — even elephants and whales!

Mammals are descendants of *cynodont therapsids,* a group of Mesozoic reptiles that have mammalian features, such as fur and specialized teeth. The transition from cynodonts to true mammals is illustrated in fossils showing subtle changes in the bone structure of the inner ear, jaw, and teeth. These creatures occupied some habitats during the Mesozoic but began to flourish only after the mass extinction of dinosaurs and other reptiles at the end of the Cretaceous period (about 65 million years ago).

REMEMBER

Three different types of mammals exist:

>> **Monotreme mammals:** Those that lay eggs instead of giving birth to live young. Modern examples include the duck-billed platypus and spiny anteater that live in Australia.

>> **Marsupials:** Mammals whose young are carried in a pouch until fully developed. Modern examples of marsupials include kangaroos and opossums.

REMEMBER

>> **Placental mammals:** By far the most successful mammal group. Most modern mammals are placental mammals, which evolved during the early Cenozoic. Placental mammals have an organ called the *placenta* that provides nourishment in the womb so that the live offspring are fairly developed by the time they are born.

CAUSING GLOBAL COOLING

A relatively recent trend among earth scientists is to view the earth's multiple systems as parts of one large planetary system. An example is found in the *uplift hypothesis* linking cooling climate patterns during the Cenozoic with the erosion of sediments from the Himalayan Mountains.

You may know from recent news about modern global warming that adding carbon dioxide gas to the atmosphere increases global temperatures by thickening the layer of greenhouse gases that surrounds the planet. Earlier in Earth's history when the climate was much warmer, higher (than present) levels of carbon dioxide and other gases existed in the atmosphere. In order for temperatures to cool down as dramatically as they did in the mid-Cenozoic, something must have removed large amounts of carbon dioxide from the atmosphere.

One way carbon dioxide is removed from the atmosphere is for it to be consumed during a chemical reaction with exposed rock materials. The carbon dioxide gas first combines with rainwater in the atmosphere. When the rain falls on exposed rocks, the carbon dioxide links itself to elements in the rock minerals (forming carbonate minerals), removing them and thus chemically weathering the rock. When this process occurs, the carbon dioxide gas is no longer in the atmosphere; it is part of Earth's lithosphere. The longer this process continues, the less carbon dioxide gas is left in the atmosphere.

The uplift hypothesis proposes that during the Cenozoic, when the Himalayan Mountains were formed or uplifted by the collision of the Indian continent into the Asian continent, massive amounts of rock were exposed, and the rate of chemical weathering increased. That increase removed large amounts of carbon dioxide gas from the atmosphere, resulting in dramatically cooler worldwide temperatures.

This hypothesis is compelling in the way that it links the cycles of Earth's lithosphere, atmosphere, and hydrosphere, but it is far from proven. Scientists who study chemical weathering, atmospheric gas concentrations, and climate change (to name a few subjects) are hard at work testing the uplift hypothesis in hopes of one day having a greater understanding of how the various systems of Planet Earth interact.

Regulating body temperature

REMEMBER

A huge advantage that mammals have over reptiles and other cold-blooded or *exothermic* animals is the ability to control and regulate their own body temperature. This ability allows them to live in environments with fluctuating or extreme temperatures because their bodies will burn or conserve energy as needed to keep a steady temperature.

This advantage, however, comes with costs. Endothermic animals must gather and store energy that can be used to warm them up. This means they need to eat more calories and more often than exothermic animals. They also have to be skilled at foraging or hunting to meet the energy demands of their endothermic body system.

Filling every niche

In ecology, a *niche* is defined as an organism's position in the ecosystem relative to other organisms. For example, some organisms (such as plants) are producers, whereas other organisms are consumers (anything that eats the plants). Mass extinctions (which I cover in the next chapter) leave many open niches for the surviving organisms to fill. Successful animals evolve and diversify to fill these niches, spreading into new environments and adapting new lifestyles.

Mammals during the Paleogene period experienced a surge in new species types as they diversified to fill niches left open by the extinction of dinosaurs. Some were *insectivores* (eating only insects), whereas others were *herbivores* (eating plants) or *carnivores* (eating other animals). The diversity of mammals in the Paleogene laid the foundation for the vast diversity of modern mammals.

Living Large: Massive Mammals Then and Now

During the Eocene epoch, which ran from about 56 to 34 million years ago, the first large land mammals appeared. This now-extinct group is called the *uin-tatheres*. An example is the giant, rhinoceros-like creature called *Uintatherium* illustrated in Figure 21-4.

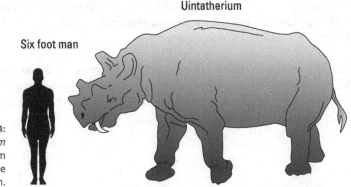

Uintatherium

Six foot man

FIGURE 21-4:
A *Uintatherium*
mammal from
the Eocene
epoch.

Although the uintatheres are extinct, ever since they appeared some type of large mammal has existed on Earth. In this section, I walk you through the well-documented evolutionary history of the two largest modern mammals, elephants and whales, and briefly describe the large mammals of the Pleistocene epoch ice ages.

Nosing around elephant evolution

Modern elephants are currently the largest land mammals. Only three species exist today, but throughout most of the Cenozoic era (up until about 2 million years ago), many more species thrived.

Elephants didn't start out so big. Elephants and their relatives are called *proboscids*, named after their large *proboscis*, or trunk. The earliest known member of this group is *Moeritherium*, which was only about the size of a modern day pig. It may have been aquatic (like a hippopotamus) and otherwise didn't look very much like an elephant. Figure 21-5 shows what a Moeritherium probably looked like.

Further up the elephant family tree is a group called *Gomphotherium*. These animals looked much more like modern elephants. One big difference was that unlike modern elephant species, which have tusks only on their upper jaws, the *Gomphotherium* had tusks growing from both their upper and lower jaws. The lower tusks were shovel-shaped, which scientists believe were used to dig up plants to eat.

The most recognizable elephant ancestors are the *mammoth* (pictured in the color photo section) and the *mastodon* of the more recent Quaternary period (beginning 2.8 million years ago). The mammoth is more closely related to modern elephant species, whereas the mastodon belongs to a more distant branch of the proboscid family tree. Both mammoths and mastodons roamed through North America until about 13,000 years ago when they, and many other large mammals, went extinct. (More on this most recent mass extinction in Chapter 22.)

A key difference between mammoths and mastodons was the size and shape of their teeth. The flat, grinding teeth of a mammoth and more cone-shaped teeth of a mastodon are illustrated in Figure 21-6.

These differences in teeth indicate that mammoths were adapted to eat grass, whereas mastodons probably ate twigs and leaves from shrubs and trees. The teeth of modern elephant species are flat, for grinding grasses, very similar to mammoth teeth.

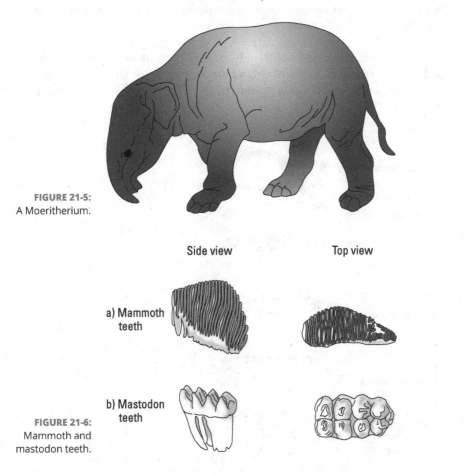

Moeritherium, an elephant ancestor

FIGURE 21-5:
A Moeritherium.

Side view   Top view

a) Mammoth
teeth

b) Mastodon
teeth

FIGURE 21-6:
Mammoth and
mastodon teeth.

Returning to the sea: Whales

Although elephants are the largest modern animals on land, blue whales are the largest animal ever — larger even than dinosaurs. Not all whales are as large as the blue whale, but they all share an evolutionary history. And the story of whale evolution does not begin in the ocean. As with all other mammals, the story of whales begins on land.

About 50 million years ago, a land-living mammal existed that scientists call *Pakicetus*. It had some traits in common with land mammals and some in common with later whale fossils. In particular, characteristics of its ear structure seem to indicate a transition between other land mammals and later aquatic mammals (whales).

Scientists are still working to fill in the details of whale evolution through fossil and genetic research. Some of the whale ancestors are shown in Figure 21-7 illustrating the physical developments in the transition from land mammals to modern marine mammals.

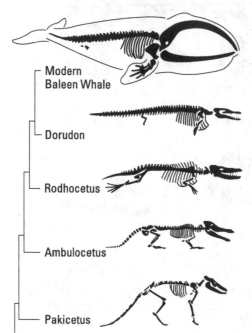

FIGURE 21-7: Stages in whale evolution from land-dwelling to fully aquatic marine mammals.

Modern Baleen Whale

Dorudon

Rodhocetus

Ambulocetus

Pakicetus

Larger than life: Giant mammals of the ice ages

Most modern mammals are somewhere in between the size of an elephant and a mouse. During the early Quaternary ice ages, however, many species of mammals had grown unusually large. These large mammals are called *megafauna*.

Along with the mastodons and mammoths, there were also very large species of camel, elk, bison, ground sloths, bears, beavers, rhinoceros, and even kangaroos. The skeletons of some of these creatures have been preserved in places like the La Brea tar pits in California and frozen into the *permafrost* (frozen soil) of Alaska, Canada, and Siberia. Preserved in this manner, soft tissue, skin, and fur, as well as bones, are recovered to be studied by scientists.

Scientists speculate that the large size of these animals may have been an adaptation to the colder climates; larger animals are able to stay warm more easily than smaller ones because they can generate more heat than they lose. Regardless, the reign of giant mammals had only just begun when it was brought to an end around 13,000 years ago.

Exactly why the ice age megafauna went extinct is still being explored, but some scientists propose that after humans showed up, the conditions of existence on this planet were changed forever.

Right Here, Right Now: The Reign of Homo Sapiens

The Cenozoic era has not ended. Many geologists consider right now to be part of the Quaternary period that began about 2.8 million years ago. Other scientists have proposed that a new period has begun: the *Anthropocene*, the geologic period of humans.

Biological classification groups humans, or *Homo sapiens*, with primates, including monkeys, lemurs, and apes. Within the primates, humans are more closely related to gorillas and chimpanzees, but they stand alone as hominids.

The evolutionary story of hominids began nearly 7 million years ago at the end of the Miocene epoch when hominids and chimpanzees diverged from a common ancestor. Fossils indicate that by 4 million years ago, australopithecines appeared, mostly in Africa. Australopithecines included five different species of hominids that were similar to modern humans in that they were *bipedal* (walking upright on two legs), but they still had many ape-like and primitive primate features. A few species of australopithecines were living side by side with the early ancestors of modern humans, *Homo habilis* and *Homo erectus*, around 2.8 million years ago.

The direct path from *Homo erectus* and *Homo habilis* to modern humans (*Homo sapiens)* is not precisely documented. Scientists have proposed and continue to test two prominent hypotheses:

>> **Out of Africa:** The "out of Africa" hypothesis suggests that early humans evolved from a single mother who lived in Africa and whose offspring migrated out of Africa around 100,000 years ago into Europe and Asia.

>> **Multiregional:** The multiregional hypothesis proposes that multiple populations of early humans were scattered across Europe and Asia. Occasional contact and interbreeding led to the characteristics of modern humans among populations far enough separated to result in the wide diversity of features we see today.

SEEKING ANSWERS TO MYSTERIES OF THE PAST

Just thinking about pollen makes some people sneeze. But pollen grains are a very useful tool to scientists asking questions about the past. Pollen grains are made of a special organic protein called *sporopollenin* that can last for millions of years if preserved under the right conditions. And pollen grains are tiny (microscopic) and abundant (millions of them fit on the tip of your pinky finger). Pollen grains come from flowering plants, and those plants need certain conditions of temperature, sunlight, water, and nutrients to survive. Finally, many species of flowering plants produces a pollen grain as distinct as its flowers — with visible, identifiable characteristics to tell a scientist exactly what plant it came from.

Scientists find pollen grains preserved in the bottoms of lakes or other sediments where they are deposited chronologically, season by season, year after year. By identifying the different plants the pollen must have come from, the scientists can answer questions about past environments such as:

- What plants were living thousands (or millions) of years ago?

- What were the conditions of temperature and moisture?

By looking at preserved pollen grains, they can also ask and answer questions about how human societies (after they appeared) affected the environments they lived in. For example, finding a change in the pollen grains from trees (indicating forest) to grass (indicating agriculture).

The next time pollen fills your eyes and nose, remember how useful those millions of irritating grains will be to future scientists seeking answer to mysteries of the past!

After arriving in Europe and Asia, a subspecies of *Homo sapiens*, the Neanderthals, dominated the landscape from about 200,000 to 30,000 years ago. *Homo sapiens neanderthalensis* had skulls very different from modern humans — a prominent brow ridge and receding chin — and were generally more muscular and shorter than modern humans. Neanderthals were, however, very similar to prehistoric humans in the way they lived: building shelters, using tools, and burying their dead with personal items and flowers.

Evidence exists that by 30,000 years ago, *Homo sapiens* existed in Europe. The rest, as they say, is history (not geology).

Arguing for the Anthropocene

Some scientists believe that we are now living in a new new geological epoch, the *Anthropocene* — the age in which human activities have a geological impact on the earth. They propose that humans, unlike the animals before them, are shaping and changing the earth in ways that are comparable to the natural processes of geology. In particular, humans act as surface shapers — shifting rock and sediment around the earth's surface. But some human activities reach deep into the earth's crust to extract resources, including water. Humans create change that previously only natural, physical processes of the earth could have created. Perhaps, indeed, the earth is experiencing changes unique to a new geologic period: the Anthropocene.

Altering the climate

One of the first topics that comes up when discussing the impact of human activities on the earth is climate change. Humans burn fossil fuels such as coal and oil for energy, releasing carbon dioxide gas into the atmosphere. Carbon dioxide and other greenhouse gases insulate the earth from the sun's harmful radiation but also trap heat that would otherwise leave Earth and radiate into the universe. Here, I briefly list some of the effects that modern climate change is having on the earth's geologic system:

>> **Increasing chemical erosion:** Adding carbon dioxide gas to the atmosphere through the burning of fossil fuels (such as coal and oil) increases the rates of chemical weathering. The interaction of carbon dioxide with exposed and uplifted rocks gradually wears away the minerals in a process of chemical weathering erosion.

>> **Killing coral reefs:** The additional carbon dioxide in the atmosphere also dissolves into the oceans, increasing their acidity. When an ocean becomes acidic, creatures that build coral reefs suffer. When ocean conditions are too acidic, coral reefs are not built and the organisms that build them and live in them are threatened by extinction.

>> **Melting glaciers and ice caps:** Increasing levels of greenhouse gasses in the atmosphere trap the earth's heat, raising global temperatures. With higher temperatures, the ice sheets and glaciers of the world begin to melt faster and are not replenished each season because the temperatures are too warm.

>> **Shifting weather patterns:** Higher temperatures in the atmosphere mean higher temperatures in the oceans as well. The interaction between air and ocean temperatures is what drives water into the atmosphere through evaporation (forming clouds) and produces air circulation patterns (wind). By increasing global temperatures, humans will dramatically change the patterns of seasonal weather that have been long established. Regions that are dry are likely to become drier and experience droughts, whereas regions that are tropical are likely to experience extensive flooding. Most of all, weather patterns will become increasingly stormy and unpredictable.

Shaping the landscape

Humans exert huge amounts of change on the surface of the earth to make it a more habitable (or profitable) place to live. The changes made to the landscape by humans may be visible to future geologists studying the Anthropocene — these changes will be evident in the ways sediment is deposited differently, in the way river channels and beaches have been straightened, dug out, or filled in, and in the removal of entire mountaintops.

Damming rivers

Along many of the world's rivers, humans have constructed dams. The dams use the flow of water to produce electricity. To do so, the natural flow of river water is slowed, creating a lake on the upriver side of the dam. Water is released through the dam as needed to produce electricity. The entire process creates a significant deviation from the natural pattern of streamflow that occurred before the dam was built.

Multiple geologic changes result from dam-building. Slowing the flow of water by building a dam also slows (or stops) the flow of sediment. Instead of being transported downstream to the ocean, the sediments are deposited in the dam-created lake. And, of course, this lake has been created where there was not a lake before, flooding habitats and displacing creatures, including (in many cases) humans.

Shaping waterways

Damming is only one way humans act as geologic agents to flowing water. Humans also straighten and deepen river channels to provide access to inland cities for the shipping and tourism industries.

Many river channels are not naturally deep enough to allow the passage of large freighter ships that transport goods around the world. The solution, of course, is to dig the river channel deeper by removing sediments. This process of *dredging* acts as a short-term, massive erosional force — taking sediments from the river channel and transporting them elsewhere.

Similarly, a river's natural path may be changed by the construction of a concrete channel to help direct the water where humans want it to go, rather than where the natural shape of the landscape sends it. The Seine River in Paris is shaped in this way — straightened into a smooth curve through the center of the city and guaranteed not to change course.

Nourishing beaches

Coastal beaches depend on rivers to carry sediments from the continent to the sea. As the river washes the sediments into the ocean, the waves carry sand up and onto the nearby beaches. But this natural geologic process of beach building is halted by changes upstream.

When humans change the course or sediment load of a river, the sediment that normally would have been brought downstream to the ocean and deposited along beaches is no longer available. To fix this problem, humans add sand to the coast in what is called *beach nourishment*. To build up beaches, sediments are brought in from somewhere else and deposited along the beachfront. However, these added sediments may not be the right size to withstand the erosional forces of waves or to support the natural ecosystem.

Changing coastlines

Another way humans act as geologic agents along the coast is by building seawalls and manmade *spits* (linear deposits of sand) to direct the flow of water and sediments in order to minimize the removal of beach sand by wave action. The goal is to maintain a beachfront for the homes and other structures that have been built. However, these structures actually reduce the natural beachside drift of sediments. In some cases, the man-made structures also concentrate and magnify wave energy — resulting in increased erosion of beach sediments and leading to the need for more beach nourishment, not less.

Removing mountaintops

In regions such as the Appalachian Mountains of North America, coal deposits are found in layers of rock forming the mountainous landscape. The coal mining industry, with the aid of huge land-moving machinery, has developed ways to extract as much coal as possible — rather than mine in underground tunnels, it simply removes the overlying layers of rock. In other words, the key is to remove the top of a mountain.

This process of coal extraction begins with removing all the trees and vegetation. Then each layer of non-coal-bearing rock is scraped away. After the layer of coal itself is stripped away, the material that has been removed may be replaced, or it may not. Often the removed materials are left in nearby valleys, polluting rivers and changing the shape of the landscape. After all, when a mountain of rock and sediment is removed, it must be deposited somewhere else.

Leaving evidence in the rock record

Perhaps the strongest argument for a new geologic era marking how humans have changed the earth and affected the rock record can be found in looking at deep sea sediments. In recent years, cities and towns have made efforts to battle plastic pollution — but it is largely too late. Plastic waste has been found in the deepest of sea trenches and the tops of mountains, from the equator to the poles. Much of this plastic waste is made of microplastics — pieces of plastic too tiny to see without a microscope. Microplastics don't look like obvious pollution the way a plastic grocery bag along the side of the road does. But they are clearly evidence of humankind changing things here on Planet Earth. Larger pieces of plastic, especially fishing lines and other elements of industrial fishing waste, can be found floating in the oceans — all of the oceans. But the microplastics are joining the other sediments on Earth, on the seafloor, in rivers, and on glacier ice, to become part of the sedimentary rock record. Plastic, being a man-made material, does not biodegrade like organic matter and does not dissolve or disaggregate like minerals under the right conditions.

If geologists in the future look to the rocks, they will see a period in Earth's history when the dominant biological organism (humans) pulled fossil fuels from deep in the earth and used them to alter the climate and create a new type of material — plastic, which will be forever evident in the layers of rock.

IN THIS CHAPTER

» Looking for causes of mass extinctions

» Finding out about five major extinctions in Earth's history

» Focusing on extinctions in human history

Chapter 22

And Then There Were None: Major Extinction Events in Earth's History

More than once in Earth's long history, geologic events have led to the demise of multiple species. Sometimes whole families of organisms disappeared, putting an end to that particular path of evolution and leaving room for surviving animals to spread into new habitats. Each of these extinctions is well-documented by changes in the fossils preserved in the geologic record.

To be *extinct* means to no longer exist. Technically speaking, at the end of your lifetime you will be extinct, though your species *(Homo sapiens)* will not be extinct because many other people will still be living. When scientists talk about extinction, they talk about the extinction of every member of a whole species, or even a *genera* (group of species) or a *family* (group of genera).

REMEMBER

When the term *mass extinction* is used, it indicates that a very large number of species cease to exist. In geologic time, a mass extinction event may occur over several hundreds of thousands — or even millions — of years. While this seems like a long time relative to a human lifespan, remember that geologically speaking it's not very long at all.

In this chapter, I explain the most common theories for what leads to mass extinctions and describe the five biggest extinction events in Earth's history, as well as a few minor ones.

Explaining Extinctions

Each mass extinction in Earth's history has been recorded by the sudden absence of fossils of certain organisms in the geological record. These events or periods of extinction affected the entire planet. Scientists think that in each case, some change in the environment resulted in conditions that could no longer support the organisms that had adapted to it. Thus, the organisms died off in great numbers, and some never reappeared.

Scientists have not yet determined, unquestionably, what led to each mass extinction, but they have some good ideas, expressed as theories, that are still being tested by modern scientific research. I introduce four such theories in this section.

Heads up! Astronomical impacts

Earth is only one of many objects moving through the universe. Occasionally, as evidenced by the craters on the moon, flying objects in space may hit one another. When this occurs, it is called an *impact event.* Scientists have found evidence for impacts on Earth, such as craters resulting from meteorites that have hit Earth's surface. The span of human history has not recorded any impact large enough to cause a dramatic change in global conditions, but evidence exists that such major events occurred in the past.

While it may seem obvious that being struck by a meteor devastates life in the areas surrounding the impact zone, what is not so obvious is the continued after-effects that are experienced all around the globe. The following sequence of events explains how global ecosystems could be negatively affected through an impact event:

1. **A large object hits Earth.** The impact sends large amounts of rock and other collision debris into the atmosphere, and it starts fires, which add smoke and ash to the atmosphere.

2. **The atmosphere is polluted.** The particles of ash and rock in the atmosphere do three things:

   - Block sunlight, which plant life depends on

   - Block sun warmth, leading to global cooling

   - Create conditions for acid rain

This darkened atmosphere may also be very cold and difficult to breathe in —
like a day of heavy smog in modern cities, but a day that lasts for many years.

3. **Plant life is affected first.** The combination of acid rain, cooler temperatures,
   and absence of sunlight shuts down the process of photosynthesis and brings
   plant life to a halt.

4. **Herbivores are affected next.** Without the plants to support them, herbivo-
   rous animal species begin to suffer.

5. **The entire ecosystem collapses.** As the plants and herbivores disappear,
   animals that depend on them (carnivores) also suffer. Eventually entire food
   webs have been affected and begin to collapse.

Keep in mind that mass extinction does not occur in one day. The sequence of
events following an impact may continue for many hundreds or thousands of
years following the impact event itself. The species that can't adapt to a new way
of life will die out.

Lava, lava everywhere: Volcanic eruptions and flood basalts

Basalt rocks formed by the cooling of lava indicate that at times in Earth's past,
volcanic activity occurred on a massive scale. Entire regions of the continents, called
provinces, are covered by layers of basalt rock many miles deep. Regions of the modern continents that are covered in these flood basalts are illustrated in Figure 22-1.

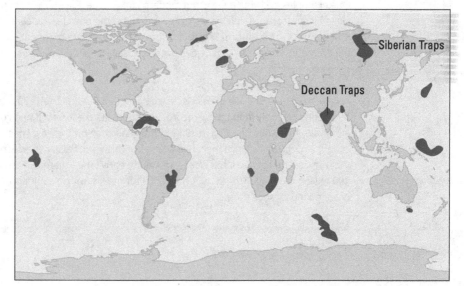

FIGURE 22-1:
Regions of the
modern
continents
covered in flood
basalt rock layers.

Such provinces are not formed by the eruption of lava from a volcanic mountain such as Mount Saint Helens, pictured in this book's color photo section, but rather from *fissures:* elongated cracks where the magma below erupts onto the surface without explosive force. Today such *fissure eruptions* are most common on the flanks of volcanic mountains and in the Hawaiian Island volcanic eruptions.

Geologists conclude that the eruption of lava from giant fissures created the flood basalts and would have also altered the global environment. Specifically, such massive eruptions of lava, much larger than the fissures currently erupting on the Hawaiian Islands, would have been accompanied by the release of huge amounts of volcanic gas into the atmosphere. The result would have been rising global temperatures and associated changes in climate patterns due to the added sulfur dioxide and water vapor in the air.

Some of these flood basalt events are thought to have lasted for hundreds of thousands of years at a time. While the region affected by the lava itself would be confined to a particular continent, the global effects of changed atmosphere and climate would have reached every part of the earth, both on land and in the oceans.

LIVING FOSSILS

For the most part, species today are physically very different from their distant ancestors. However, a special few organisms alive today look exactly like their ancestors from millions of years ago. These species are called *living fossils.*

One example of a living fossil is the modern ginkgo tree. Fossils indicate that the earliest ginkgo trees from 170 million years ago had the same physical appearance as ginkgoes living today. Certain internal characteristics of the tree have changed, but compared to the dramatic changes in other trees over the last 170 million years, it appears the ginkgo hasn't evolved much at all.

Another example of a living fossil is the coelacanth fish. Until the 1930s, the coelacanth had been seen only in the fossil record, and scientists presumed it had gone extinct along with the dinosaurs. But in 1938, some fishermen off the east coast of South Africa caught a coelacanth. Since then, two species of this living fossil have been identified in that region. The coelacanth is closely related to early lobe-finned fish that evolved 400 million years ago (check out Chapter 19 for details on fish evolution) and hasn't changed much since then.

Additional examples of living fossils include crocodiles, some species of shark, and the giant panda.

Shifting sea levels

During certain periods in Earth's history, most of life was lived in shallow oceans. A shift in sea level would have had dramatic effects on the environments supporting shallow marine life. Lower sea levels would force shallow sea life into dry, waterless environments. Higher sea levels would leave them in deeper water with less access to sunlight and oxygen found near the ocean's surface.

Such sea-level changes could have been the result of climate changes (the melting or growing of large ice caps, which would change the amount of water in the oceans) or tectonic plate movements (which I describe in Chapter 9).

Changing climate

Most scientists now consider climate change to be the most important factor in mass extinctions. The earth's climate changes in response to many different factors, including impacts, tectonic plate movements, and volcanic eruptions.

In looking at evidence for mass extinctions in the geologic records, scientists conclude that global-scale changes can most reliably be explained through changes in a global system, such as the climate. Other geologic evidence indicates that periods of mass extinction commonly occur during global warming or glaciations, leading scientists to conclude that shifting climate conditions changed global environments so dramatically that many species could not adapt, and perished.

End Times, at Least Five Times

Species go extinct all the time; extinction is part of the natural order of things. Normal rates of extinction through time are part of what scientists call the *background extinction rate* expected to occur on Earth. The mass extinctions described in this section are periods when the rate of extinction, as indicated in the fossil record, is much more dramatic and extreme than the normal (background) rate.

Figure 22-2 is a graph that illustrates extinction rates throughout Earth's history, highlighting the five major extinction events that I describe in this section.

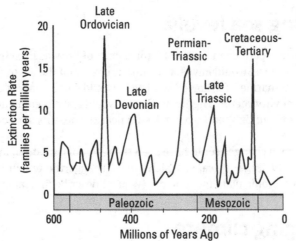

FIGURE 22-2:
Extinction rates
for five major
extinction events.

Cooling tropical waters

The first major mass extinction that scientists know about happened approximately 445 million years ago toward the end of the Ordovician period (in the Paleozoic era). At the time, life was lived in the oceans; no evidence indicates that land plants or animals existed yet. Scientists think the expansive marine environment was affected by a cooling climate and abrupt changes in sea level as extensive glaciers grew over the continents of the South Pole. More than 100 families of marine organisms, primarily those living in tropical regions near the equator, went extinct. This totaled more than 50 percent of the living families of that period.

Scientists conclude from the evidence for glaciation that the colder climate at the poles meant conditions in the tropics were also cooler, leaving the warm water-adapted organisms nowhere to go. The amount of water locked up as ice over the South Pole may also have dramatically lowered sea levels all over the planet, reducing the habitat for undersea organisms.

Reducing carbon dioxide levels

At the end of the Devonian period (also in the Paleozoic era), around 370 million years ago, another extinction event affected marine life. This event seems to have affected reef-building organisms living in shallow marine environments, as well as some groups of early land plants.

Only slightly less dramatic than the earlier extinction event, the Late Devonian extinction saw almost 50 percent of the existing families disappear. The fact that organisms in shallow marine waters as well as on land were affected has led scientists to hypothesize that atmospheric conditions, such as changes in carbon dioxide levels, played a large role in this event.

The early plants themselves may have altered atmospheric levels of carbon dioxide through photosynthesis. Less carbon dioxide leads to cooler global climate conditions, which may then have affected marine life in warm, shallow sea ecosystems.

The Great Dying

A mass extinction event that marks the transition between the Paleozoic and Mesozoic eras, about 250 million years ago, is called *The Great Dying*, the *Permian-Triassic event*, the *Permo-Triassic extinction*, or the *End-Permian extinction*. At this time, more than 96 percent of species in the oceans and 70 percent of the species on land (including some plants) perished. The End-Permian extinction is the only extinction event in Earth's history to affect insects, resulting in a loss of 33 percent of the insect species of the time.

Scientists are not certain what caused the End-Permian extinction. This event appears to have occurred over a few million years, leading scientists to rule out an impact as the primary cause.

At the time of this extinction, the supercontinent of Pangaea (see Chapter 19) was forming, which may have changed ocean circulation patterns and temperatures more quickly than species could adapt to. But some scientists argue that by the time of the extinctions, the landmasses had already moved and shouldn't have further changed the marine environments in any important way.

This mass extinction was most severe in the oceans, leading some scientists to conclude that global water conditions must have experienced a disruptive change. One explanation may be that the oceans became *anoxic*: lacking in dissolved oxygen.

TECHNICAL STUFF

Oxygen levels in the ocean are maintained by the circulation of surface waters, which cool near the poles, sink (taking oxygen-rich water into the deep sea), and move back toward the equator. This circulation of water due to changes in temperature and salinity (the amount of salt it contains) is today called the *thermohaline ocean conveyor*.

Scientists propose that toward the end of the Permian period (the last period of the Paleozoic era), climate conditions were so warm all over the planet that ocean circulation was stopped — with the result that no oxygen was brought into the deep sea, which essentially suffocated marine life.

On the continents, the *Siberian Traps* — a massive outflowing of lava from volcanic fissure activity in what is now northern Siberia — likely affected global climate conditions. The release of gases associated with this type of volcanic activity may have been enough to create a global greenhouse, warming temperatures enough to halt the temperature-driven cycling of oxygen in the oceans.

Paving the way for dinosaurs

At the end of the Triassic period (the first period of the Mesozoic era), about 200 million years ago, approximately 35 percent of the animal families became extinct. While this event is the least dramatic of the five major extinctions, its cause is also still a mystery to scientists.

This extinction event was likely spread over a long period of time. The *Central Atlantic magmatic province* — a region of massive lava flows between the continents of South America, Africa, and Europe as Pangaea split apart — was erupting, and evidence for climate conditions suggests the climate was on a roller coaster from one extreme to the next. This unsteady climate may have made it difficult for some species to adapt, resulting in their extinction.

The end-Triassic extinctions paved the way for dinosaur dominance. As other animal groups died out, dinosaurs expanded to fill the empty niches, eventually covering every environment on Earth.

Demolishing dinosaurs: The K/T boundary

Possibly the most well-known mass extinction event is the one that ended the reign of the dinosaurs at the end of the Cretaceous period (at the end of the Mesozoic era). In the geologic record, the transition from the Cretaceous period to the Paleogene period is well-marked by the disappearance of dinosaur fossils. As I explain in Chapter 20, some dinosaurs are the ancestors of birds. These avian dinosaurs are the only ones who survived this extinction event. All the reptilian dinosaurs are found in rock layers below this time period's geologic layer — not above it.

The geologic layer marking the boundary between the Cretaceous and Paleogene periods is called the *K/T boundary*. The K stands for the German word for Cretaceous, *Kreidezeit*; the T is for *Tertiary*, which is the period between the Cretaceous and the Quaternary (65 to 2.8 million years ago). Modern geologists who no longer recognize the Tertiary period have begun to refer to this transition as the *K-Pg boundary* (Cretaceous-Paleogene) instead.

Many events occurred during this time that may have, together, resulted in the extinction of so many animals. The supercontinent of Pangaea was breaking up, and the Deccan Traps of India (see Figure 22-1) were erupting. As I explain earlier in the chapter, tectonic plate movements, as well as massive volcanic activity, can change global atmospheric, climate, and ocean conditions, resulting in animal extinctions.

At the K/T (or K-Pg) boundary, however, there is also clear evidence of an impact event. In the Gulf of Mexico, just off the north side of the Yucatan Peninsula, scientists have identified a massive crater. The *Chicxulub Crater* illustrated in Figure 22-3 is the result of a rocky body at least 10 kilometers (6 miles) wide hitting the earth around 65 million years ago.

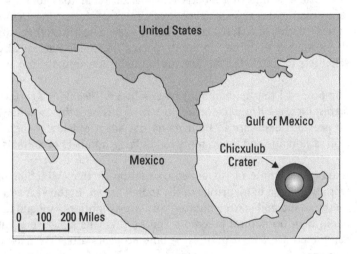

FIGURE 22-3: The location of the Chicxulub crater in the Gulf of Mexico.

According to the impact theory (which I describe earlier in this chapter), such an impact could have created long-lasting darkening of the atmosphere, interfering with plants first, and then rippling up the food chain to devastate the largest creatures: dinosaurs.

The strongest line of evidence for this explanation is the amount of iridium found in layers of sediment dating to this time. *Iridium* is an element that is rare in Earth's crust but much more common in meteors. Its worldwide presence in layers of clay and sand that must've been at the surface of the earth during the K-Pg boundary indicates that somehow, large amounts of iridium were introduced to Earth's atmosphere. Scientists accept that a large meteor impact is an obvious explanation.

Modern Extinctions and Biodiversity

In this section, I describe a significant extinction event in the age of man. After humans evolved and began to spread across the globe, the large mammals of the Cenozoic era (which I describe in Chapter 21) began to decline. In this section, I present possible explanations for that decline, and I touch on ideas about how man's continuing impact on our planet may affect biodiversity.

Hunting the megafauna

About 14,000 years ago, humans first entered the Americas, probably by way of a land bridge connecting Siberia and Alaska. What they found was a land full of large mammals, or *megafauna*, such as mammoths, mastodon, bison, horses, ground sloths, and rhinoceros to name a few. Shortly after the arrival of humans, the numbers of large mammal species declined dramatically, leading some scientists to conclude that humans hunted these animals into extinction.

The proposal that human hunting resulted in megafauna extinction is called the *prehistoric overkill hypothesis.* Supporters of this hypothesis claim that animals with no previous exposure to humans do not adapt quickly enough to the predatory skill of humans and their ability to kill large numbers of animals at once.

A recent example of an overkill situation is the extinction of the dodo bird from the island of Mauritius in the Indian Ocean. In the year 1600, sailors arrived on the island and started hunting this large flightless bird and its eggs to eat. The dodo, with no natural predators and no previous experience with humans, was completely wiped out by 1681 — only 80 years after it started co-existing with humans.

Supporters of the overkill hypothesis also point to similar events in Australia more than 40,000 years ago. When human populations first arrived in Australia, numerous species of large *marsupials* (kangaroo-like mammals) went extinct. The cause of this extinction — like the North American megafaunal extinctions — is still being debated. Some scientists think it was a direct result of human migration to the continent and overhunting of the animals. Other scientists suggest that human changes to the environment (through the use of fire to clear vegetation) were a more important factor in causing the extinctions.

However, other hypotheses have been proposed to explain the disappearance of these large mammals:

>> **Asteroid impact:** Recently, scientists have presented evidence for a possible impact event that may have led to the extinction of North American megafauna.

Researchers are still hotly debating this possibility and looking for evidence (such as where such an impact may have occurred) to support their hypotheses.

>> **Climate change:** Other scientists claim that climate changes occurring at the same time, such as ice age glaciers melting and global conditions becoming warmer, were significant enough to lead some species into extinction. However, the climate change hypothesis leaves skeptics wondering why large mammal species didn't migrate to new habitats as the environments changed. And others suggest that the climate didn't change quickly or dramatically enough to result in extinctions.

Reducing biodiversity

The human effect on megafauna, or large mammals, is far from over. Today, many of the animals listed as threatened or endangered by humans are the largest existing mammals, including the American bison, Asian elephants, mountain gorillas, various species of whales, and many species of bear. As humans continue to dominate the earth, the pattern of humans moving into new regions and leading species to extinction is ongoing.

Human population growth and expansion into new ecosystems threatens many species with extinction. While a background rate of extinction is normal and expected, some *ecologists* (scientists who study ecosystems) suggest that humans have increased the extinction rate by up to 10,000 times its normal rate.

The existence of multiple species is called *biological diversity* or *biodiversity*. Scientists realize that biodiversity is very important because ecosystems with high biodiversity are more likely to adapt in response to disturbances (such as wildfires). Richly biodiverse regions like the rainforests of South America are also home to species of plants, insects, and animals that may have important and undiscovered medicinal properties.

The most damaging effect humans have on biodiversity is the destruction, fragmentation, and pollution of ecosystems. Many of the most biodiverse regions are also the most fragile and are easily damaged by the building of roads, introduction of non-native species (or farm animals), industrial pollution, and deforestation. If these species become extinct, whether due to human-caused climate change or development and land use, humans may lose something of great value that they didn't even know they had!

6

The Part of Tens

Discover ten ways that you use geologic resources in your everyday life, from the gemstones in your jewelry, the gypsum in your walls, and the precious metals used to build your smartphone.

Find out ten geologic hazards to be aware of, the most common ways geology destroys human endeavors and sometimes takes human lives.

Chapter **23**

Ten Ways You Use Geologic Resources Every Day

Throughout this book you have learned about rocks found on Earth and the processes that create them. Maybe your understanding of geology is still of something that happens "out there" away from where you are comfortably sitting, reading this book. In this section, I highlight a few of the most common ways that you use geologic resources. This includes in your home, in your neighborhood, and even in your computer!

Burning Fossil Fuels

One of the best known and most pervasive ways that humans use geologic resources is through the use of fossil fuels: coal, oil, and natural gas, to create energy. The hydrocarbons that are extracted from deep in the earth are the remains of once-living organisms (no, not dinosaurs; most are more likely from algae of the distant past). As ancient organic material is trapped under layers of clay and sand it slowly decays, but without access to oxygen it cannot fully decompose. Instead the hydrocarbons molecules just hang out. Depending on the conditions of burial and the location, they form different types of fossil fuel resources that humans can collect. Some organic matter gets crushed over time, turning from a peat bog into bituminous coal that can be mined, such as what is extracted in the Appalachian Mountains of North America. Other organic matter may be trapped in the seafloor sediments and over millions of years form a liquid that can be drilled out as oil, such as that found in the Gulf of Mexico seafloor. Tar sands, such as found in Canada, are sandy sediments with sticky hydrocarbon tar filling in all the spaces between grains of sand. And natural gas that is captured by fracking is simply gaseous hydrocarbons hanging out in the tiny spaces between sediments in clay, shale, and sandstone rocks.

All of these different types of "fossil fuel" release carbon dioxide when burned to provide energy and with recent understanding of how CO_2 effects climate change, there has been a push to move away from our global dependence on these resources.

Playing with Plastics

Plastics are man-made products that were engineered out of the leftover molecules from processing fossil fuels like liquid oil and petroleum. Since their invention in the early 1900's, plastics have become the most common material for making almost everything! Unfortunately because plastics are man-made, they do not recycle back into the ecosystem the way naturally occurring products do, like paper (from wood fiber) or metals. Look around your house and count all the items made of plastic. These count as geologic resources too, because without fossil fuels, there would be no plastics.

Gathering Gemstones

For millennia humans have gathered beautiful minerals, called gemstones, to adorn themselves with as jewelry. Many of the common gemstones, such as those assigned as birth stones to each month of year, are simply common minerals with either eye-catching color or intriguing characteristics of hardness. For example, some minerals of quartz have minor amounts of other elements trapped in their crystal structure. When this happens you get a gemstone! When quartz has manganese trapped within it, it turns purple and we call it an amethyst. When an amethyst is heated, its color changes to yellow and we call it a citrine. Similarly, both sapphire and ruby are the mineral corundum, but the sapphires have minor amounts of titanium present and the rubies have minor amounts of chromium present. Other common gemstone minerals include emeralds, aquamarines, and opals.

Drinking Water

Maybe you haven't thought of water as a geologic resource before, but without rock layers to filter groundwater, the water in your well wouldn't be as fresh! As water moves across the land surface it also moves underneath the land surface (see Chapter 12). The movement of water through the tiny pore spaces of permeable rock helps filter and clean the water. By the time it reaches an aquifer where it can be stored and accessed, it may already be clean enough to drink straight from the well. However, fresh groundwater is an endangered resource. Rising sea levels threaten to infiltrate fresh water aquifers in coastal regions, which, once contaminated by saltwater, are no longer useful for human consumption. Another danger to the groundwater resources in many parts of the world is fracking for natural gas. While the fracking doesn't directly contaminate groundwater, fracking does release gasses such as methane that have been trapped in the rocks along with other natural gasses. Once these gasses are released during the fracking process, they can't be contained and kept from moving through the rocks into nearby freshwater aquifers.

Creating Concrete

Maybe you thought the concrete that is used to build sidewalks, buildings, bridges, and other structures in your city or town was another type of rock. Or maybe you haven't thought about it at all. The truth is that concrete is a man-made rocklike material. To make concrete a number of different rocks are actually used. Firstly you need some crushed rocks and stones and some water. But to glue them together you need cement — or mineral glue. In nature, sedimentary rocks are glued together or cemented with minerals such as calcite or quartz that precipitate between the sediment grains and glue them into one large piece. To make cement humans have mimicked this process by combining limestone, clay, and gypsum (another precipitating mineral). Once you have the crushed rocks, water, and cement mixture, stir it together and let sit until dry. Now you have concrete!

Concrete is the workhorse material for human buildings. It is used in creating the foundation to homes and buildings, as well as shaping the landscape with sidewalks, highways, concrete barriers, and other structures.

Paving Roads

Roadways criss-cross almost every landscape now, a sure sign of the Age of Humans or Anthropocene (see Chapter 21). To build these roads humans depend on a number of different geologic resources to create asphalt, that black stuff that roads are made of. Asphalt is a combination of sticky tar, called bitumen or bituminous coal. The asphalt can be mined where it formed naturally in the rock layers, or refined from other extracted fossil fuels. Combined with gravels, sand, and a concrete mix, the asphalt serves as the sticky glue to hold these materials together where they are laid down and compressed to create a roadway.

Accessing Geothermal Heat

As our understanding of climate change and the effect of additional carbon in the atmosphere has grown, many regions have looked to alternative sources of energy. One of these sources is geothermal heat. As you dig deeper into the earth you will experience an increase in heat that scientists called the geothermal gradient. The depth at which heat from inside the earth is accessible varies with location. Regions near volcanic activity in particular have access to a great amount of geothermal heat at relatively shallow depth beneath the surface.

To access this heat, for example, to heat your home, you need a system of pipes built that extends deep underground below your house. By sending water through theses pipes, the water will be heated in the deep sections simply because the pipes are surrounded by heated rocks. After it's heated, the water will be circulated back up into your home where it can be used for a heat source instead of using gas, oil, or other fossil fuels to generate heat.

Fertilizing with Phosphate

Phosphate is an important mineral to support life and is found in the rocks of Earth's surface. Natural processes bring phosphate to surface environments through erosion of rocks as they are uplifted on Earth's surface. Phosphate is also mined and refined into fertilizers for agricultural use. The expanding agricultural business of the last century has ramped up the mining and application of phosphates — the mineral helps plants grow and when applied to crops can increase yields dramatically. However, much of the phosphate used is not absorbed by plants and ends up washing into nearby waterways, which can become overloaded with the mineral, leading to negative side effects such as excessive algae blooms.

Constructing Computers

Computers and handheld device technology depend heavily on geologic resources. You can start by examining the plastics in the device shell or casing, but the real story is on the inside. Computers are built using a huge variety of elements mined from minerals found all over the world. Here is a few of the elements found in certain parts of your computer or smartphone:

>> **Screen display:** Displays are commonly made of materials containing lead and tin, as well as quartz and potassium or lead for the glass. The lead for strengthening the glass screens is mined from the mineral galena.

>> **Circuit board:** The internal workings of a computer or smart device depend on a long list of metals and other elements mined from minerals, including but not limited to: silicon from quartz, copper from chalcopyrite, and aluminum from bauxite.

>> **Function:** Certain functions of your smartphone depend on the magnets inside fueled by the rare-earth element neodymium. Without this vital resource, your phone would not be able to run its vibration motor for notifications.

Building with Beautiful Stone

The rocks we find on Earth are beautiful with their wide array of minerals, textures, and colors. And while the foundation of your home may be the man-made concrete, the many attractive accents or outer layer may be adorned with rock selected not for its strength, but simply for its beauty.

Using attractive stone for construction and decoration began in ancient times in many societies, and the tradition continues today. Although the front of many homes and building may have natural stone features, or accessories, in recent decades there has been a boom in using polished granites, marbles, quartzites, and other rock types for interior decoration as well. Countertops of cut stone are popular, and if you shop around you will find slabs of interesting and beautiful stone in many different colors to choose from. Similarly, interesting stone is used for flowing tiles and even in showers and bathrooms as tile.

Chapter **24**

Ten Geologic Hazards

L iving on Earth can be dangerous. Earth is a dynamic planet, geologic processes are always in motion, and while some processes take millions of years (like the formation of a sedimentary rock), others happen very quickly (like a volcanic eruption). All these processes are natural, but they become hazards when they affect human lives and the infrastructure of modern society. In this chapter, I describe some common and not-so-common geologic hazards.

Changing Course: River Flooding

All over the world, water flows from higher elevations to lower elevations as stream or river flow. The movement of water through a region over a long period of time carves out riverbeds and stream channels. But rivers don't always stay in the same place. Under the right conditions, they may change course or overflow their banks, resulting in a flood.

Floods are caused by many factors, but the bottom line is that there is too much water in one place at one time and the streamflow of the river can't move it downstream quickly enough. This event is very common in the springtime, as rainfall melts snow and overloads the stream channels with water. As the water spills over the banks of the river channel, whole towns may find themselves inundated.

WARNING

Lives lost to flooding are often the result of *flash flooding* — when rainfall occurs so quickly and heavily that there is no warning before water comes rushing across roadways and taking out bridges. Drivers and their vehicles may be swept away in a current of flood water. Another dangerous situation during floods is when people attempt to drive through pools of standing water, not realizing how deep it really is until it's too late.

Caving In: Sinkholes

Sinkholes occur when groundwater erodes underground rock, removing the bedrock material that is supporting the land surface, so that the soil, sediment, and any structures built on it collapse. Sinkholes can occur anywhere that the underlying rocks have water-soluble minerals in them and are most common in regions with karst topography (which I describe in Chapter 12). In such regions, the underground erosion creates caves and cavities beneath the surface, so there's not enough material to support the land surface above.

Sinkholes often cave in right after a heavy rainfall; the added weight of the water in the soil is too heavy for the roof of the underground cavity to support, and it collapses. When a sinkhole forms, collapsing inward, it leaves a funnel-shaped depression on the ground surface. Anything on that surface sinks into the hole, including roads, bridges, buildings, and homes.

Sliding Down: Landslides

Landslides are a type of mass wasting (see Chapter 11) where large amounts of rock and soil move downslope rapidly. They are most hazardous in populated and developed regions, where they may destroy homes or block roads.

Landslide conditions are created under a number of different circumstances that result in unstable slopes — where the materials on the slope cannot be supported, and everything slides down. The potential for a landslide increases when heavy storms bring rain that adds weight to the land surface or when earth-shaking events such as earthquakes or volcanoes occur. Landslide conditions can also be created by erosion (a topic I discuss later in this chapter).

Shaking Things Up: Earthquakes

Earthquakes occur all over the planet all the time. But they become geologic hazards when they occur near the surface, are particularly strong, or happen in regions that are heavily populated. The combination of these factors is what determines the damage and death toll of an earthquake event.

REMEMBER

The amount of damage resulting from an earthquake is related not only to its strength and location but also to the type of structures it affects. In some regions, such as the state of California, measures have been taken to construct buildings that can withstand the occasional ground-shaking activity of earthquakes. As seen in the 2010 earthquakes that affected Haiti and Pakistan, less modernized regions may be devastated by an earthquake.

When earthquakes occur, they may initiate a phenomenon called *liquefaction.* Liquefaction is when the shaking movement of the earthquake leads saturated sediments (commonly sand) to flow as if they are a liquid. The liquefaction of sediments by earthquakes can result in all the damages associated with landslides.

Efforts to record and study earthquakes are designed to eventually predict them in hopes of avoiding the destruction of property and loss of human life that often result from earthquake activity.

Washing Away Coastal Towns: Tsunamis

Another hazard, a tsunami, is closely related to earthquakes and landslides. When either of these two events occurs underwater, it creates a large wave that can travel hundreds of miles an hour across the open ocean, washing ashore as a wall of water and rushing miles inland. Sometimes incorrectly referred to as *tidal waves* (they have nothing to do with the tides), these giant waves have been recorded in South America as a result of earthquakes that occurred in Japan.

Similarly, offshore, undersea landslides can displace large amounts of water, sending the water onshore with devastating outcomes. Coastal towns affected by landslide-generated tsunamis often have little to no warning.

REMEMBER

Earthquakes do provide some warning that a tsunami may be on its way. If you are near the coast and feel an earthquake, or if you see the water move far out to sea, get to higher ground immediately. (The water often moves out, like a very fast, low tide, before the large tsunami waves come ashore.)

Damages from tsunamis include the destruction of property and vegetation from the fast-flowing flood waters, as well as the inland water flow or *inundation* into areas relatively far from shore. Since the 2004 Indian Ocean tsunami, much research on tsunamis (and their relation to earthquakes) has centered on predicting these events and building warning systems to notify communities that may be in danger when a tsunami occurs. Unfortunately, as the 2011 tsunami in Japan illustrated, the warning systems are most effective for regions that are far enough away to respond to warnings before being struck by the massive waves.

Destroying Farmland and Coastal Bluffs: Erosion

Compared to a volcanic eruption, tsunami, or earthquake, erosion may seem like nothing to worry about. The truth is that you are more likely to be affected in your daily life by soil erosion than you are by any of the more dramatic geologic hazards.

Erosion is the movement of sediments and soil due to wind, water, or other processes. Erosion of soil is most common in areas that are cleared of vegetation (because plants hold the soil in place), such as by overgrazing of herd animals or deforestation. After the soil is removed, the land can no longer support vegetation — or crops — and it becomes useless.

WARNING

This situation is particularly hazardous in hot, dry regions where cropland, food, and water are already scarce. The erosion of surface soils in such regions has increased with warming global temperatures in the last century: The higher temperatures reduce rainfall, leaving surface sediments more susceptible to removal by wind and making life much more difficult for the people who live there.

Erosion from moving water, such as along rivers, streambeds, and coastlines, is also a hazard and can lead to landslides by undercutting the angle of repose (see Chapter 12). As the bluffs along rivers or the coast are eroded, any buildings sitting on them are in danger of collapsing into the water below.

Fiery Explosions of Molten Rock: Volcanic Eruptions

As geologic hazards go, volcanic eruptions make for good photo ops, with fiery red lava and large ash clouds flowing from mountaintops. The nature of a volcanic eruption and, thus, the hazards it presents vary depending on the type of volcano and the composition of the molten rock inside. Some volcanic eruptions include slow flowing lava pouring out, while others are explosive, sending massive amounts of gas, ash, and broken rock materials miles into the air and down mountainsides.

REMEMBER

While the lava that flows from volcanoes is extremely hot and dangerous, it moves slowly and doesn't present an immediate danger to nearby towns. The real hazards from volcanic eruptions are in the form of ash clouds and pyroclastic flows. Ash clouds from volcanoes can travels for miles in the air. The ash can block the sun, reduce visibility for drivers, and damage airplane engines. *Pyroclastic flows* are debris flows of intensely heated rock and ash that move down mountains through the valleys. Unlike flowing lava, pyroclastic flows move very quickly (almost 450 mph) and destroy everything in their path.

And finally, volcanic eruptions can release large amounts of poisonous gases into the atmosphere, which endanger the health of people in nearby communities.

Melting Ice with Fire: Jokulhlaups

The Icelandic term *jokulhlaup* (sometimes pronounced *yer-kul-hyolp*) refers to a flood from sudden melting of a glacier. In Iceland, where glaciers cover a volcanic landscape, small jokulhlaups often occur when the ice is heated by volcanic activity, melts, and rushes to the sea. These "glacial outburst floods" cause all the damage associated with massive flooding and occur with the unpredictability of a volcano.

Jokulhlaups, while first named in Iceland, are not restricted to Iceland. Any region where glaciers form is in danger of experiencing a jokulhlaup. A jokulhlaup can occur when a glacier that dams a lake shrinks enough that the lake water spills out suddenly. The eruption of glacier-covered volcanoes, such as Mount Rainier, Washington, would result in jokulhlaup-type flooding, as well as debris flows from the combination of meltwater and volcanic debris produced during the eruption.

Flowing Rivers of Mud: Lahars

When large amounts of water are mixed with sediments, they produce a flowing river of mud called a *lahar*. Lahars often occur with other hazards covered in this chapter, including jokulhlaups and volcanoes. In the case of lahars from volcanic activity, the volcanic eruption of heated rock and lava can melt glaciers and snow, resulting in lahars. Jokulhlaups, similarly, may result in lahars when the melted glacial water mixes with loose, muddy sediments.

As lahars move through towns, they flow like water but with much greater force. When they stop moving, everything they've inundated becomes cemented in the drying mud.

Watching the Poles: Geomagnetism

The earth's magnetic field may not seem like much of a hazard. On the contrary, it is quite useful if you are lost in the woods with a good compass. Occasionally, however, the strength of the earth's magnetic field changes, growing stronger or more active. These periods are called *magnetic storms*, and during magnetic storms the earth's magnetic field may interfere with technological systems that humans depend on. GPS systems and other satellites can be disrupted. The changes in magnetic activity can also cause blackouts when they affect electrical grids. To prepare for the potential hazards of geomagnetism, the U.S. Geological Survey (USGS) works with NASA and the U.S. Department of Defense to monitor the earth's magnetic field activity.

Index

atomic structure, 52–56

atomic symbol, 54

atoms, 52–56

  chemical bonds, 56–60

    covalent bonds, 57–58

    ionic bonds, 57

    metallic bonds, 58–60

  formulating compounds, 60

  ions, 56

  isotopes, 56

  molecules, 56

  periodic table, 53–55

aureole, 106

avalanches, 174

B

background extinction rate, 337–341

backswamps, 187

backwash, 225

banded iron formations (BIFs), 276, 277

bar graph, 20

barchan dunes, 218–219

basal sliding, 198

basalt, 81

  flood, 335–336

  vs. gabbro, 24–25

  growth of, 63

  at hot spots, 156

  observing, 88

base level, 183

basins, 103, 143–144

batholith, 93

bathymetry, 124

baymouth bar, 229

beaches, 229, 331

bed load, 181, 213

bedding planes, 103

bedrock, 167, 168

beds, 103

beneficial mutation, 256

benthic foraminifera, 303

beta capture, 242

beta decay, 242

BIFs (banded iron formations), 276, 277

bifurcating pattern, 262

biodiversity, 342–343

biogenic chert, 101, 102

biogenic limestone, 101, 102

biogenic sedimentary rocks, 101

biological classification, 274–275

biological diversity, 343

biosphere, 40

biostratigraphic units, 237

bipedal, 310, 327

birds, Mesozoic era, 313–314

bituminous coal, 102

black shale, 100

black smokers, 277

blowouts, 215

blue-green algae, 272

body fossils, 14, 259

body plan, 282

body temperatures, mammals, 322–323

body wave, 160

bony fish, 289

boulders, 100, 206

Bowen, Norman L., 81–82

Bowen's reaction series, 81–83

bracketing, 249

braided streams, 185–186

breakers, 225

breccia, 100, 101

brittle deformation, 142

brittle minerals, 66

brooks. *See* streams

Burgess Shale, 283

burial metamorphism, 106

burrows, 260

C

C-14 dating, 248

Cal atomic symbol, 54

Calamites, 291–292

calcite, 74, 75, 101

calcium element, 54

calderas, 89, 157

calving, 197

Cambrian Explosion, 14, 278–279, 282–284

capacity, stream, 181

capillary action, 190

graded beds, 104
graded streams, 183
gradient, streams, 180
Grand Canyon, 319
granite-gneiss complexes, 268
gravel, 100
gravity, 12
  cause of mass wasting, 165, 169
  and friction, 166
  law of, 22
  theory of, 22
The Great Dying, 339
greenstone belts, 268–269
Grenville Orogeny, 271
greywacke sandstones, 100
ground surface, 188
groundwater, 177, 188–194
  flow of, 190–192
  infiltrating tiny spaces, 188
  karst, caves, and sinkholes, 192–194
  measuring porosity and permeability, 189
  water table, 189–190
group number, 53
guide fossils, 285
gymnosperms, 292
gypsum, 75, 101
gypsum evaporites, 102

H

hadrosaurs, 311
halite, 68–69, 101
hanging valleys, 201
hanging wall, 145
hardness, of minerals, 64–65
headlands, 228
heat, 150–515
heat transfer melting, 80
heavy metal, 44
Heezen, Bruce, 33, 124
hematite, 76
herbivores, 310, 323
Hess, Harry, 33
high altitudes/latitudes, 196

high tide, 226
Homo erectus, 327
Homo habilis, 327
Homo sapiens, 327–329
horizontality, 30
hornfels, metamorphic minerals, 110, 112
horns, 202
horsts, 147
hot rock, decompressing, 80
hot water vents, 276
hotspots, 154–157
humus, 94
Hutton, James, 30
Hutton's hypothesis, 30–31
hydraulic gradient, 189
hydraulic lifting, 182
hydrocarbons, 347–348
hydrologic cycle, 40, 176–178
hydrolysis, 97
hydrosphere, 40
hydrothermal metamorphism, 106
hydrothermal vents, 276, 277
Hypocrosaurus, 311
hypothesis, 8
  to answer questions, 16
  continental drift, 32, 118–124
    equatorial supercontinent, 123
    evidence of, 118–119
    fossil matching, 119–120
    glacier ice movement, 122
    search for mechanism of, 123–124
  Giant Impact, 266
  Hutton's, 30–31
  multiregional, 327
  out of Africa, 327
  predictions based on, 16
  prehistoric overkill, 342
  scientific method, 18–19
  snowball earth, 34, 278
  solar nebula, 266–267
  uplift, 322
  using methodical approach, 18–19

I

ice, 12. *See also* water
  caps, 196, 330
  flows on glaciers, 196–199
  movement on glaciers, 122
  sheets, 177, 196, 202
  shelves, 196
ice age, 206
  cycling, 207
  giant mammals during Cenozoic era, 326–327
ice sheet glacial erosion, 202
icebergs, 197
ichnofossils, 260
ichthyosaurs, 307
igneous rocks, 9, 79–93
  below surface, 92–93
  Bowen's reaction series, 81–83
  classifying, 85–88
  evolving magmas, 83–84
  formed when crystallized, 84–85
  from magmas, 80
  melt composition, 81
  studying volcanic structures, 89–92
  transforming, 111
impact event, 334
impermeable layers, 191
impurities
  gemstones, 76
  minerals, 64
inclusions, 238
index fossils, 285
indirect pressure, 108–109
infiltration, 188
inheritance of acquired traits, 254
inner core, 44
inorganic chert, 101
inorganic limestone, 101, 102
inorganic minerals, 62
inorganic sediment grains, 101
insectivores, 323
insects, 305
intercontinental seaway, 302
interglacials, 207, 320
intermediate minerals, 81
intermediate rocks, 85

internal deformation, 198
internal temperatures, 269–270
interpreting results, scientific method, 21
intracontinental basin, 103
intrusive igneous rocks, 84
invertebrate, 284
ionic bonds, 57
ions, 56
Iridium, 341
iron element, 54
iron-rich minerals, 81
irregular minerals, 66
island arcs, 155
isostasy, 132–133
isostatic equilibrium, 132–133
isostatic rebound, 209–210
isotopes, 13, 56, 241

J

joints, 146
jokulhlaups, 357
Juan de Fuca Plate, 302, 318
Jurassic period, climate, 301

K

K atomic symbol, 54
kames, 204–206
kaolinite group, 74
karst, 192–194
kettle lake, 205
kimberlite pipes, 77
komatiites, 269–270
K/Pg boundary, 340–341
K/T boundary, 340–341

L

laccolith, 93
lag deposit, 221
lag feature, 221
lahars, 172, 174, 357–358
lakes, 177
  kettle, 205
  oxbow, 185–186
  paternoster, 201–203

About the Author

Alecia M. Spooner teaches at Seattle Central College, where she is Professor of Earth and Environmental Sciences. She has earned degrees in anthropology (B.A. University of Mississippi), archaeology (M.A. Washington State University), and geology (M.S. University of Washington). Her research included interdisciplinary studies of paleoecology and archaeology using fossil pollen. These days, she teaches earth science courses that are accessible and engaging while stressing scientific literacy and critical thinking.

Author's Acknowledgments

I want to acknowledge and thank the following people: my friend, partner, and love Chris, who knows when to keep me grounded and when to encourage me to soar; my three kids, who share their wonder with me every day; my colleagues, peer mentors, and students, who've made me the teacher I am; and my agent, along with the team at Wiley, who know a good idea when they hear one.

Dedication

For my students who are constantly teaching me new ways to ask questions and offering solutions to problems I didn't realize I had.

"Nothing in life is to be feared, it is only to be understood. Now is the time to understand more, so that we may fear less."

— MARIE CURIE

Publisher's Acknowledgments

Acquisitions Editor: Ashley Coffey
Project Editor: Christopher Morris
Copy Editor: Christopher Morris
Production Editor: Siddique Shaik
Proofreader: Debbye Butler

Cover Image: © Manuel Breva Colmeiro/ Getty Images

Take dummies with you everywhere you go!

Whether you are excited about e-books, want more from the web, must have your mobile apps, or are swept up in social media, dummies makes everything easier.

Find us online!

Leverage the power

Dummies is the global leader in the reference category and one of the most trusted and highly regarded brands in the world. No longer just focused on books, customers now have access to the dummies content they need in the format they want. Together we'll craft a solution that engages your customers, stands out from the competition, and helps you meet your goals.

Advertising & Sponsorships

Connect with an engaged audience on a powerful multimedia site, and position your message alongside expert how-to content. Dummies.com is a one-stop shop for free, online information and know-how curated by a team of experts.

- Targeted ads
- Video
- Email Marketing
- Microsites
- Sweepstakes sponsorship

20 MILLION PAGE VIEWS EVERY SINGLE MONTH

15 MILLION UNIQUE VISITORS PER MONTH

43% OF ALL VISITORS ACCESS THE SITE VIA THEIR MOBILE DEVICES

700,000 NEWSLETTER SUBSCRIPTIONS TO THE INBOXES OF *300,000* UNIQUE INDIVIDUALS EVERY WEEK

# of dummies

Custom Publishing

Reach a global audience in any language by creating a solution that will differentiate you from competitors, amplify your message, and encourage customers to make a buying decision.

- Apps
- Books
- eBooks
- Video
- Audio
- Webinars

Brand Licensing & Content

Leverage the strength of the world's most popular reference brand to reach new audiences and channels of distribution.

For more information, visit dummies.com/biz

PERSONAL ENRICHMENT

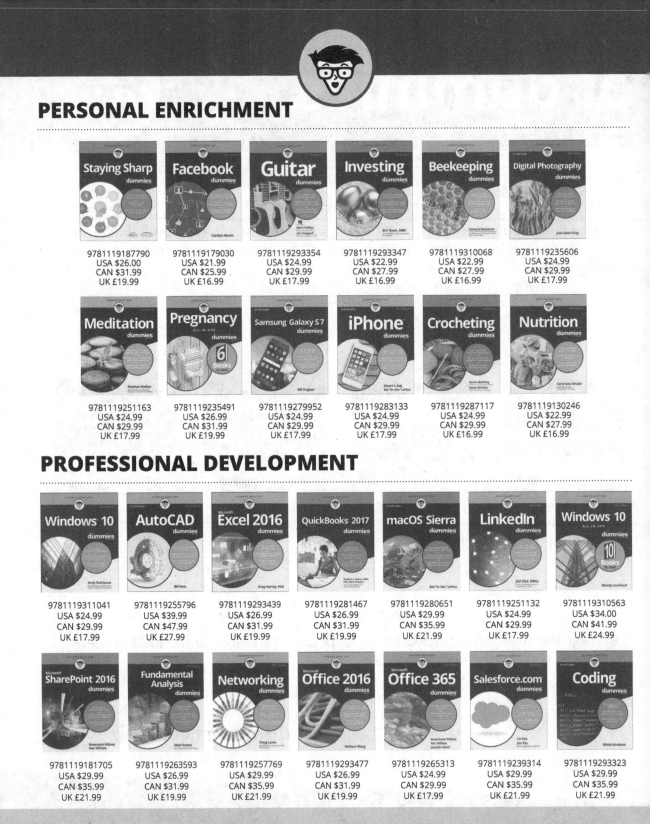

Staying Sharp dummies
9781119187790
USA $26.00
CAN $31.99
UK £19.99

Facebook dummies
Carolyn Abram
9781119179030
USA $21.99
CAN $25.99
UK £16.99

Guitar dummies
Mark Phillips
Jon Chappell
9781119293354
USA $24.99
CAN $29.99
UK £17.99

Investing dummies
Eric Tyson, MBA
9781119293347
USA $22.99
CAN $27.99
UK £16.99

Beekeeping dummies
Howland Blackiston
9781119310068
USA $22.99
CAN $27.99
UK £16.99

Digital Photography dummies
Julie Adair King
9781119235606
USA $24.99
CAN $29.99
UK £17.99

Meditation dummies
Stephan Bodian
9781119251163
USA $24.99
CAN $29.99
UK £17.99

Pregnancy ALL-IN-ONE dummies
9781119235491
USA $26.99
CAN $31.99
UK £19.99

Samsung Galaxy S7 dummies
Bill Hughes
9781119279952
USA $24.99
CAN $29.99
UK £17.99

iPhone dummies
Edward C. Baig
Bob "Dr. Mac" LeVitus
9781119283133
USA $24.99
CAN $29.99
UK £17.99

Crocheting dummies
Karen Manthey
Susan Brittain
9781119287117
USA $24.99
CAN $29.99
UK £16.99

Nutrition dummies
Carol Ann Rinzler
9781119130246
USA $22.99
CAN $27.99
UK £16.99

PROFESSIONAL DEVELOPMENT

Windows 10 dummies
Andy Rathbone
9781119311041
USA $24.99
CAN $29.99
UK £17.99

AutoCAD dummies
Bill Fane
9781119255796
USA $39.99
CAN $47.99
UK £27.99

Excel 2016 dummies
Greg Harvey, PhD
9781119293439
USA $26.99
CAN $31.99
UK £19.99

QuickBooks 2017 dummies
Stephen L. Nelson, MBA, CPA, MS in Taxation
9781119281467
USA $26.99
CAN $31.99
UK £19.99

macOS Sierra dummies
Bob "Dr. Mac" LeVitus
9781119280651
USA $29.99
CAN $35.99
UK £21.99

LinkedIn dummies
Joel Elad, MBAs
9781119251132
USA $24.99
CAN $29.99
UK £17.99

Windows 10 ALL-IN-ONE dummies
Woody Leonhard
9781119310563
USA $34.00
CAN $41.99
UK £24.99

SharePoint 2016 dummies
Rosemarie Withee
Ken Withee
9781119181705
USA $29.99
CAN $35.99
UK £21.99

Fundamental Analysis dummies
Matt Krantz
9781119263593
USA $26.99
CAN $31.99
UK £19.99

Networking dummies
Doug Lowe
9781119257769
USA $29.99
CAN $35.99
UK £21.99

Office 2016 dummies
Wallace Wang
9781119293477
USA $26.99
CAN $31.99
UK £19.99

Office 365 dummies
Rosemarie Withee
Ken Withee
Jennifer Reed
9781119265313
USA $24.99
CAN $29.99
UK £17.99

Salesforce.com dummies
Liz Kao
Jon Paz
9781119239314
USA $29.99
CAN $35.99
UK £21.99

Coding dummies
Nikhil Abraham
9781119293323
USA $29.99
CAN $35.99
UK £21.99